H. G. Hornung
E.-A. Müller (Eds.)

Vortex Motion

H. G. Hornung / E.-A. Müller (Eds.)

Vortex Motion

Proceedings of a colloquium held at Goettingen on the occasion of the 75th anniversary of the Aerodynamische Versuchsanstalt in November 1982

With 109 Figures

Springer Fachmedien Wiesbaden GmbH

CIP-Kurztitelaufnahme der Deutschen Bibliothek

Vortex motion: proceedings of a colloquium held at
Göttingen on the occasion of the 75th anniversary of
the Aerodynam. Versuchsanst. in November 1982/
H. G. Hornung; E.-A. Müller (eds.).

ISBN 978-3-528-08536-0 ISBN 978-3-663-13883-9 (eBook)
DOI 10.1007/978-3-663-13883-9

NE: Hornung, Hans G. [Hrsg.]; Aerodynam.
Versuchsanstalt ⟨Göttingen⟩

Editors:

H. G. Hornung
DFVLR-AVA, Institut für Experimentelle Strömungsmechanik

E.-A. Müller
Max-Planck-Institut für Strömungsforschung

1982

ISBN 978-3-528-08536-0

P R E F A C E

The institution which, for the largest part of its existence,
was called Aerodynamische Versuchsanstalt, has experienced
several changes of name as well as changes of organizational
structure. Nevertheless, continuity of its identity has been
maintained by the fact that, across these changes, the
scientists and technicians largely remained loyal to the AVA.
Its establishment at Goettingen in November 1907, under the
leadership of Ludwig Prandtl, is an important landmark in the
history of aeronautical research. The aim of the colloquium
is to celebrate the 75th anniversary of this event.

The choice of the topic "VORTEX MOTION" is not intended to be
representative of the work of the AVA. It was chosen as one
of the topics of which experimental as well as theoretical
aspects are of wide current interest throughout fluid
mechanics. Not all of the speakers who were invited from
the international fluid mechanics community were able to
accept, so that one or two aspects of the field, that we
had hoped to be covered, cannot be presented. It is, of course,
impossible to cover the subject exhaustively with a small
number of contributions. Rather, the contributions represent
isolated features from a broad spectrum right across the
field.

The colloquium is being financed by the Deutsche Forschungs-
und Versuchsanstalt für Luft- und Raumfahrt. The Deutsche
Gesellschaft für Luft- und Raumfahrt provided organizational
support. Dr. H.-G. Knoche gave valuable assistance in the
selection of the speakers.

Goettingen, July 1982   H. G. Hornung

                          E.-A. Müller

CONTENTS

On vortex formation and interaction with solid boundaries   1
N. DIDDEN

Vortices following two-dimensional separation   18
A. DYMENT

Vortex sheets and concentrated vorticity. A variation on the   31
theme of asymptotic modelling in fluid mechanics
J. P. GUIRAUD and R. ZEYTOUNIAN

Wave propagation, instability, and breakdown of vortices   50
S. LEIBOVICH

Widerstandsreduzierung bei kraftfahrzeugähnlichen Körpern   68
H. LUDWIEG

Analogies between oscillation and rotation of bodies induced   82
or influenced by vortex shedding
H. J. LUGT

A point vortex method applied to interfacial waves   97
D. W. MOORE

Vortices in turbulence   106
A. E. PERRY, M. S. CHONG and T. T. LIM

On the relation between the thunderstorm updraft and   122
tornado formation
R. ROTUNNO

Structure and stability of streets of finite vortices   142
P. G. SAFFMAN

The representation of planar separated flow by regions of   157
uniform vorticity
J. H. B. SMITH

ON VORTEX FORMATION AND INTERACTION
WITH SOLID BOUNDARIES

by

N. Didden

Max-Planck-Institut für Strömungsforschung, Göttingen, F. R. Germany
and Department of Aerospace Engineering
University of Southern California, Los Angeles, USA.

ABSTRACT

In many cases the formation of vortices is caused by an ejection of
vortical boundary layer fluid into ambient fluid. An experimental study of
two types of this process is discussed here. The first type occurs in
an impulsively started flow at the edge of a solid boundary and is characterized by the separation and roll-up of a vortex sheet. The roll-up process
of a ring vortex is studied by flow visualization in a water tank. Measurements of the unsteady flow field by means of Laser-Doppler-Anemometry
provide data for the computation of the vorticity flux into the vortex. The
measured growth rate of the vortex circulation is compared with predictions from theoretical models.

Another type of vortex formation is associated with the unsteady
separation of a boundary layer at a flat plate induced by vortices in the
outer flow. This process involves the interaction of vortices with a solid
boundary. As an example the boundary layer of an axisymmetric wall jet
produced by an air jet (Re = 19000) impinging normally on a flat plate is
investigated. By applying periodic forcing to the jet the regularity of ring
vortices in the jet shear layer is enhanced and a phase reference is provided. Phase locked flow visualization of the wall jet boundary layer gives
evidence for a separation region which moves downstream and evolves
into a secondary vortex separating from the plate. Phase averaged hot-wire
measurements show the thickening of the boundary layer, and vorticity
contours at various stages of the separation process show the ejection of
boundary layer vorticity into the outer flow.

1. Introduction

Vortex formation frequently results from the separation of a
boundary layer from a sharp edge and subsequent roll-up of the separated
vortex sheet into a vortex spiral. For the case of a flow starting from rest
important aspects of this process are the production of vorticity in the
developing boundary layer at the edge, and the transport of vorticity into
the growing vortex spiral. An understanding of both these aspects is
essential for the prediction of the vortex circulation. Initially, as long as
the size of the vortex is small compared to length scales of the boundary,
the flow can be modelled using similarity arguments. Based on similarity
solutions of Kaden (1931) for the rolling-up of a semi-infinite vortex sheet,

Wedemeyer (1961) studied the growth rate of a vortex spiral shedding from
the edges of a flat plate in a starting flow normal to the plate. Saffman (1978)
applied these similarity solutions to the roll-up of ring vortices and could
explain some results of Liess and Didden (1976) and of Maxworthy (1977).
Blendermann (1969) and Pullin (1978) obtained more general similarity
solutions predicting the vortex shape and the circulation of a plane vortex
sheet, which separates from a wedge with arbitrary edge forming angles
and for velocity-time functions $U \sim t^m$. In the first part of this paper measurements will be presented on the formation of ring vortices generated by
fluid being ejected from a circular nozzle. Details of the measurements
were reported in Didden (1979), subsequently refered to as I. Here the
emphasis will be placed on comparing the measured vorticity flux into the
rolling-up vortex with theoretical models predicting the circulation growth
rate; in particular with the slug flow model, and with results from similarity theory calculations by Pullin (1979).

In the second part of this paper we will consider another type of
vortex formation, namely that which results from unsteady boundary layer
separation induced by a concentrated vortex moving close to a flat plate. In
this case the separation leads to local ejection of boundary layer fluid into
the outer flow and to the formation of a new, but weaker vortex. The
formation of such a secondary vortex was observed by Harvey and Perry
(1971) for trailing wingtip vortices approaching ground, and by Cerra and
Smith (1980) and Schneider (1980) for ring vortices impinging on a flat plate
normal to the direction of propagation. In a theoretical study Walker (1978)
considered the boundary layer induced by a straight vortex moving at
constant height above a flat plate and found separation slightly ahead of the
vortex. He recognized separation by the appearance of a closed recirculation cell within the boundary layer (in a vortex-fixed frame of reference),
associated with rapid boundary layer thickening.

The measurements presented here were performed with the aim of
studying the flowfield in the vicinity of a vortex-induced unsteady boundary
layer separation. In the shear layer of an axisymmetric air jet impinging
normally on a flat plate, strong ring vortices were generated by periodically forcing the jet. At the plate the impinging jet formed an axisymmetric
radially spreading wall jet. Using flow visualization, the wall jet boundary
layer was observed to separate periodically as the ring vortices in the wall
jet spread radially, moving along the plate. Slightly downstream of each
ring vortex a secondary vortex evolved from the downstream moving separation region. The phase reference from the forcing signal made it possible
to obtain phase averaged velocity profiles from hot wire measurements
performed in the vicinity of the separation region. The emphasis here will
be on localizing the separation point and identifying characteristic features
of the flow field associated with the separation process.

2. Ring Vortex Formation

2.1 Experimental Set-Up

The ring vortices were produced in a watertank (0.4 x 0.4 x 1 m) at
the open end of a pipe (diameter D = 5 cm) with wall thickness of 0.5 mm at
the nozzle edge. Fluid was ejected by a hydraulically controlled piston with

a velocity $U_p(t)$, which is best approximated by

$$U_p(t) = (U_o \cdot t_2^{-m})t^m, \qquad U_o = 4.6 \text{ cm/s}. \qquad (1)$$

After a short acceleration period (m = 1/2) for $t \leq t_2$ = 0.24 s the piston moved at constant velocity (m = 0) with fluctuations less than $0.04\,U_o$.

Axial and radial velocities u and v in x- and y-direction respectively (see figure 1) were measured using a laser-Doppler velocimeter described in Didden (1977). For measurements in the nozzle-exit plane and upstream at x ≤ 0 a glass window allowed the laser beam to penetrate through the pipe wall. The circulation Γ_v of the fully formed vortex at x_o = 3 D downstream from the nozzle was determined by integrating the velocity along the symmetry axis y = 0 and closing the integration path at infinity (u = v = 0). Since the flowfield is approximately steady in a vortex-fixed frame while it convects with velocity U_T over the measuring point, the integral is

$$\Gamma_v = - U_T \int_o^\infty u\,(x_o,\ y = 0, t)\ dt.$$

2.2 Flowfield at the Nozzle

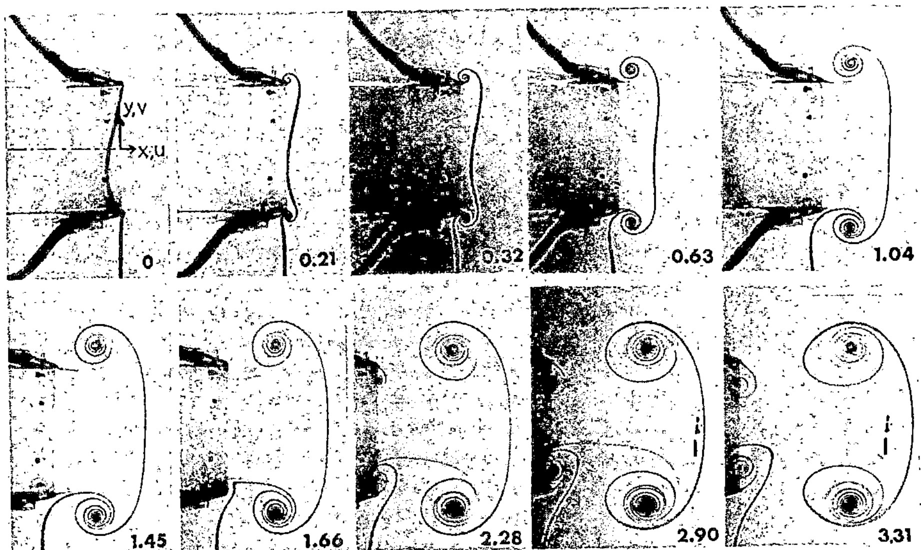

Figure 1   Side view of the rolling-up ring vortex. Visualization with dye in the vertical symmetry plane. Time in seconds. End of stroke at t = 1.6 s.

The rolling-up process of the ring vortex is visualized by dye injection in a vertical symmetry plane (figure 1). The starting flow separates at the nozzle edge and vortical fluid of the initially thin boundary layer rolls up into a spiral. From velocity measurements a smooth velocity distribution within the spiral was found, suggesting that the vorticity of the initially thin shear layer diffuses rapidly. The ring vortex moves downstream due to the self induced velocity of the circular toroidal vortex core. At the end of the piston stroke (t = 1. 6 s) the flowfield of the vortex ring produces a secondary vortex of opposite circulation at the nozzle edge. This feature is important for the estimate of convective transport of boundary layer vorticity into the rolling-up vortex since,owing to the secondary vortex flow, the vorticity flux through the nozzle exit into the vortex is stopped at the end of the piston stroke.

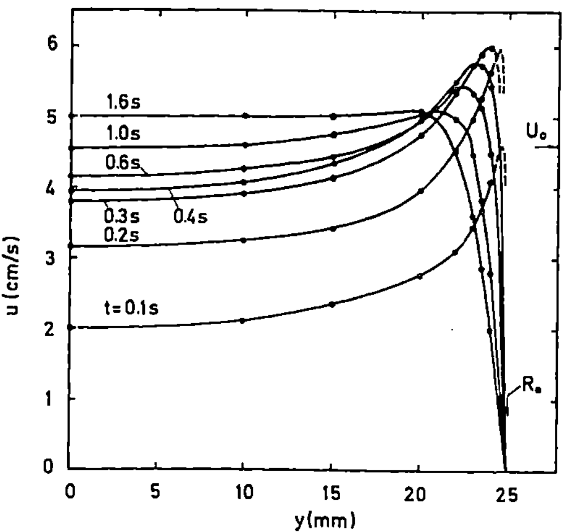

Figure 2: Profiles of axial velocity u in the nozzle-exit plane x = 0. Inner nozzle wall at y = R_0.

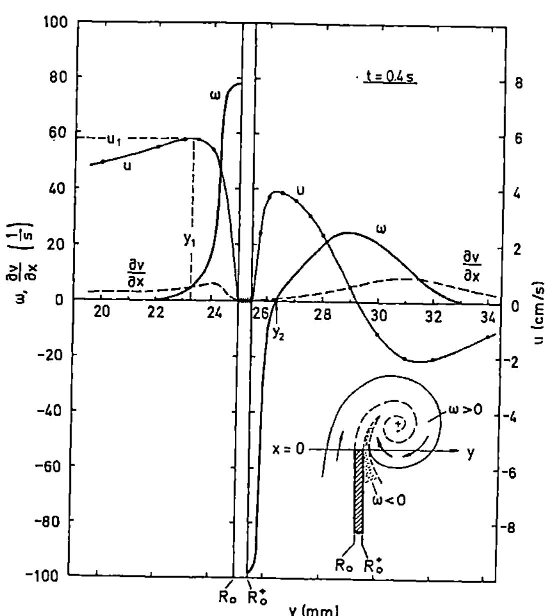

Figure 3: Profiles of velocity u and vorticity $\omega = \partial v/\partial x - \partial u/\partial y$ at x = 0 and t = 0. 4 s. The sketch shows schematically vorticity transport across x = 0.

For small times after the piston start the axial velocity profiles u (y) in the exit plane x = 0 (figure 2) exhibit a strong maximum u_1 close to the nozzle wall at R_o = 25 mm, indicating a starting flow around the nozzle edge as expected from potential theory. For a piston stroke $L/D \geq 0.2$ (or $t \geq 0.3$ s) the maximum u_1 decreases and its position receeds from the wall as the boundary layer thickness increases. For L/D = 1.4 (t = 1.6 s) the u-velocity outside the boundary layer is nearly constant across the nozzle-exit plane.

In order to calculate the vorticity flux into the rolling-up vortex as a function of time the vorticity $\omega(y, t) = \partial v/\partial x - \partial u/\partial y$ in the nozzle-exit plane x = 0 was determined from u(y) and v(y) measurements at $y = R_o$ and $y = R_o^+ = 25.5$ mm (R_o and R_o^+ are the y-coordinates of the inner and outer nozzle wall respectively). The term $\partial v/\partial x$ was evaluated from measurements of the radial velocity v(x, y, t) in the region -1mm $\leq x \leq 3$mm. As an example the profiles of u, $\partial v/\partial x$ and ω at t = 0.4 s corresponding to a piston stroke L/D = 0.28 are shown in figure 3. The vorticity of the internal boundary layer at $y \leq R_o$ is concentrated mainly between the wall and the position y_1 of the maximum velocity u_1. At the outer wall $y = R_o^+$ the flow of the rolling-up vortex produces an external boundary layer of negative vorticity, which because of u > 0 is convectively transported into the region x > 0 and rolled up into the vortex. This is illustrated schematically in figure 3. As will be shown in the next chapter, the total circulation of the ring vortex is considerably diminished by this flux of negative vorticity. The positive vorticity in the region $y > y_2$ (y_2 is the position of ω = 0) is part of the vorticity distribution within the rolling-up vortex.

2.3 Vortex Circulation

The rate of change of circulation $d\Gamma/dt$ in the region x > 0 due to convective transport of vorticity across the plane x = 0 is (see I):

$$\frac{d\Gamma}{dt} = \frac{d\Gamma_i}{dt} + \frac{d\Gamma_e^-}{dt} = \int_0^{R_o} \omega \cdot u \, dy + \int_{R_o^+}^{y_2} \omega \cdot u \, dy \qquad (2)$$

where subscripts i and e denote internal and external boundary layers. The flux of positive vorticity at $y > y_2$ (see figure 3), does not contribute to the increase of the ring vortex circulation. Thus the second term in the above equation is integrated between R_o^+ and y_2 to account only for the flux $d\Gamma_e^-/dt$ of negative vorticity of the external boundary layer. As discussed in I the integrals can be evaluated without knowledge of $\partial u/\partial y$ in the wall region $y_1 < y < y_2$, where the profiles $\partial u/\partial y$ could not be measured accurately (cf. figure 2). The total circulation $\Gamma = \Gamma_i + \Gamma_e^-$ of the fluid rolled up into the vortex ring was obtained by integrating $d\Gamma/dt$.

The time functions $d\Gamma/dt$ and Γ are plotted in figure 4. For comparison the circulation Γ_v measured at x_o = 3D downstream in fully formed vortex rings produced with different piston strokes L/D is plotted as well. The ring vortex circulation Γ_v is considerably lower than the circulation Γ_i produced by vorticity from the inner nozzle wall. This confirms that negative vorticity from the external boundary layer reduces the total circulation to $\Gamma < \Gamma_i$. The deviation of Γ from the vortex circulation Γ_v shed from the nozzle edge might be due to the fact that the vorticity flux $d\Gamma_i/dt$ initially (for $t < 0.3$ s) is sensitive to the interpolation of the u-velocity near the wall. In addition, vorticity flux across the exit plane after the end of the piston stroke could result in larger circulation Γ, but due to the formation of secondary vortices it is reasonable to assume that this vorticity flux makes no significant contribution to the vortex circulation. In the following discussion the experimental results will be compared with theoretical models for the ring vortex circulation, namely with the slug flow model, and with results from similarity theory.

2.4 Discussion

(a) Slug flow model, Γ_{SL}

For a slug flow with constant velocity $u(y, t) = U_p(t)$ over the nozzle cross-section the vorticity flux is (Didden, 1977):

$$\frac{d\Gamma_{SL}}{dt} = \frac{1}{2} U_p^2 \tag{3}$$

With equation (1) this yields $\frac{1}{2} U_0^2 \, t/t_2$ during the acceleration phase $t \leq t_2$ and $\frac{1}{2} U_0^2 = $ const. for $t \geq t_2$. Several discrepancies between the measured vorticity flux $d\Gamma_i/dt$ and the flux $\frac{1}{2} U_p^2$ derived from the slug flow model can be seen in figure 4:

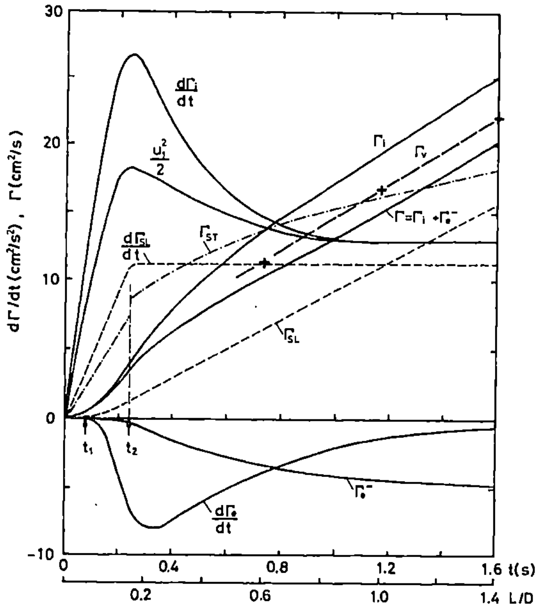

Figure 4: Circulation rate $d\Gamma/dt$ and circulation Γ versus time and corresponding piston stroke L/D (acceleration phase at $t \leq t_2$). The solid lines are experimental results utilizing eqn.2: i = internal, e = external boundary layer. Γ_v :measured circulation of fully formed ring vortices. --- SL: slug flow model eqn. 3, --·-- ST: similarity theory eqn. 5 (Pullin 1978,1979).

(i) Initially at $t < 1.0$ s or $L/D < 0.8$ the vorticity flux $d\Gamma_i/dt$ is considerably larger than $\frac{1}{2} U_p^2$ due to the large flow velocities $u > U_p$ near the nozzle edge.

(ii) At $t \geq 1.0$ s the vorticity flux $d\Gamma_i/dt$ is equal to $\frac{1}{2} u_1^2$, where u_1 is the maximum velocity of the profile $u(y)$. This is a consequence of $\partial v/\partial x \ll \partial u/\partial y$ in the boundary layer. During this phase of the vortex formation the vorticity flux is constant, in agreement with the prediction based on the slug flow model. The absolute value of the measured flux $\frac{1}{2} u_1^2$, however, is larger than predicted, since owing to the displacement effect of the boundary layer the maximum velocity u_1 remains larger than the piston velocity U_p.

(b) Similarity theory, Γ_{ST}

At the initial instant of the piston motion a potential flow around the nozzle edge is set up. In order to keep the velocity finite (Kutta condition) the boundary layer separates at the nozzle edge and forms a free shear layer rolling up into a vortex spiral. For a spiral size $d_s \ll D$ the flowfield near the edge can be derived from similarity theory by considering the potential flow starting from rest and by modelling the separated shear layer by a vortex sheet expanding from the nozzle edge into the irrotational flow. For a piston velocity $U_p \sim t^m$ Pullin (1979) derived the circulation Γ_{ST} (ST: similarity theory) shed into a plane vortex sheet. Assuming that the cylindrical vortex sheet at the nozzle edge can locally be considered a plane sheet, he found

$$\Gamma_{ST} = K_2 \left(\frac{0.75\, a^4}{1+m}\right)^{1/3} t^{4/3(1+m)-1}, \qquad a = \left(\frac{D}{2\pi}\right)^{1/2} U_p t^{-m} \quad (4)$$

with K_2 being a constant weakly dependent upon m. The axisymmetric pipe geometry has been taken into account by the geometry factor $(2\pi)^{-1/2}$ of the potential flow. For $t > t_2$ the similarity solution for $m = 0$ is expected to be valid asymptotically for a piston motion of constant velocity, having started at a virtual time t_1, which is determined by the condition $L(t) = U_o (t-t_1)$ for $t \geq t_2$. Comparison with $L(t_2)$ from time integration of equation (1) yields the virtual time $t_1 = \frac{1}{3} t_2$. From figure 12 in Pullin (1978) we find $K_2 = 2.32$ and 2.64 for $m = \frac{1}{2}$ and $m = 0$ respectively. This results in:

$$\Gamma_{ST} = 0.541 \left(\frac{D\, U_o^2}{t_2}\right)^{2/3} t \qquad\qquad \text{for } t \leq t_2 \qquad (5a)$$

$$\Gamma_{ST} = 0.704 \left(D\, U_o^2\right)^{2/3} \cdot \left(t - \frac{1}{3} t_2\right)^{1/3} \qquad \text{for } t > t_2 \qquad (5b)$$

Some important consideration can be made by comparing the vortex circulation Γ_{ST} (figure 5) with the measured circulation Γ_i, which is shed from the internal boundary layer:

(i) During the acceleration phase at $t \leq t_2$ the circulation Γ_{ST} is about a factor 2 larger than Γ_i.

The similarity calculations are expected to be valid only after an initial period Δt of slow viscous flow, i.e. after the size of the vortex spiral becomes larger than the boundary layer thickness. An estimate of Δt is obtained from the comparison of the radial distance y_w of the vortex center from the nozzle edge (see I, figure 6) with the boundary layer thickness $\delta = 4\left(\nu t\right)^{1/2}$ on a flat plate for impulsively started flow. This crude estimate yields $\Delta t = 0.20$ s, indicating that the initial period of slow viscous flow cannot be neglected.

(ii) During the constant velocity phase ($t \geq t_2$) the circulation Γ_{ST} increases much slower than Γ_i.

From flow visualization the typical diameter of the vortex spiral is found to be $d_s/D < 0.1$ for $L/D < 0.7$. Thus the assumption $d_s \ll D$ of the similarity theory is reasonably well satisfied in the range $L/D < 0.7$. The discrepancy between the growth rates of Γ_{ST} and Γ_i therefore is most likely due to the fact that the self-induced downstream motion of the ring vortex is neglected in the similarity solution for plane flow, which predicts an upstream motion of the vortex center into the region $y > R_0$, $x < 0$. This might have a major influence upon the vorticity production at the nozzle, since the flow there depends upon the position of the vortex.

(iii) For larger piston strokes a deviation of the flow from the similarity solution is expected since the spatial scale of the vortex becomes comparable with the nozzle diameter. For $L/D > 0.8$ the flow at the nozzle exit approaches a steady state with $d\Gamma/dt = $ const. , as compared to the similarity solution with $d\Gamma/dt \sim t^{-2/3}$. From figure 4 it is obvious that the agreement of Γ_{ST} with the vortex circulation Γ_v at a single point ($L/D \simeq 0.9$) cannot be used as evidence for the applicability of the similarity theory to ring vortices generated with large piston strokes $L/D > 0.8$.

In the following we summarize the deficiencies of the models, which are available for the prediction of the ring vortex circulation:

- The slug flow model neglects the unsteadiness of the starting flow with initially high velocity due to the potential flow around the nozzle edge and thus underestimates the vortex circulation. For larger piston strokes $L/D > 0.8$, when the flow in the exit plane approaches the steady state of a jet flow, the circulation growth rate is well predicted, if the displacement effect of the pipe boundary layer is taken into account.

- The similarity theory does not model the initial period of viscous flow and thus initially overestimates the circulation. Furthermore the approximation of the axisymmetric flow by a plane flow neglects the selfinduced motion of the ring vortex,which due to the curvature of the vortex core is significant in the flow regime with $d_s \ll D$ also. It is probably for this reason that the growth rate of the vortex circulation predicted by similarity calculations of Pullin (1979) was found to be smaller than the measured growth rate.

3. The Vortex-Induced Boundary Layer Separation

3.1 Jet Facility and Flow Visualization

The air jet apparatus (figure 5a) with a nozzle of D = 3.8 cm diameter was operated at an exit velocity $U_0 = 7.5$ m/s, corresponding to a Reynolds

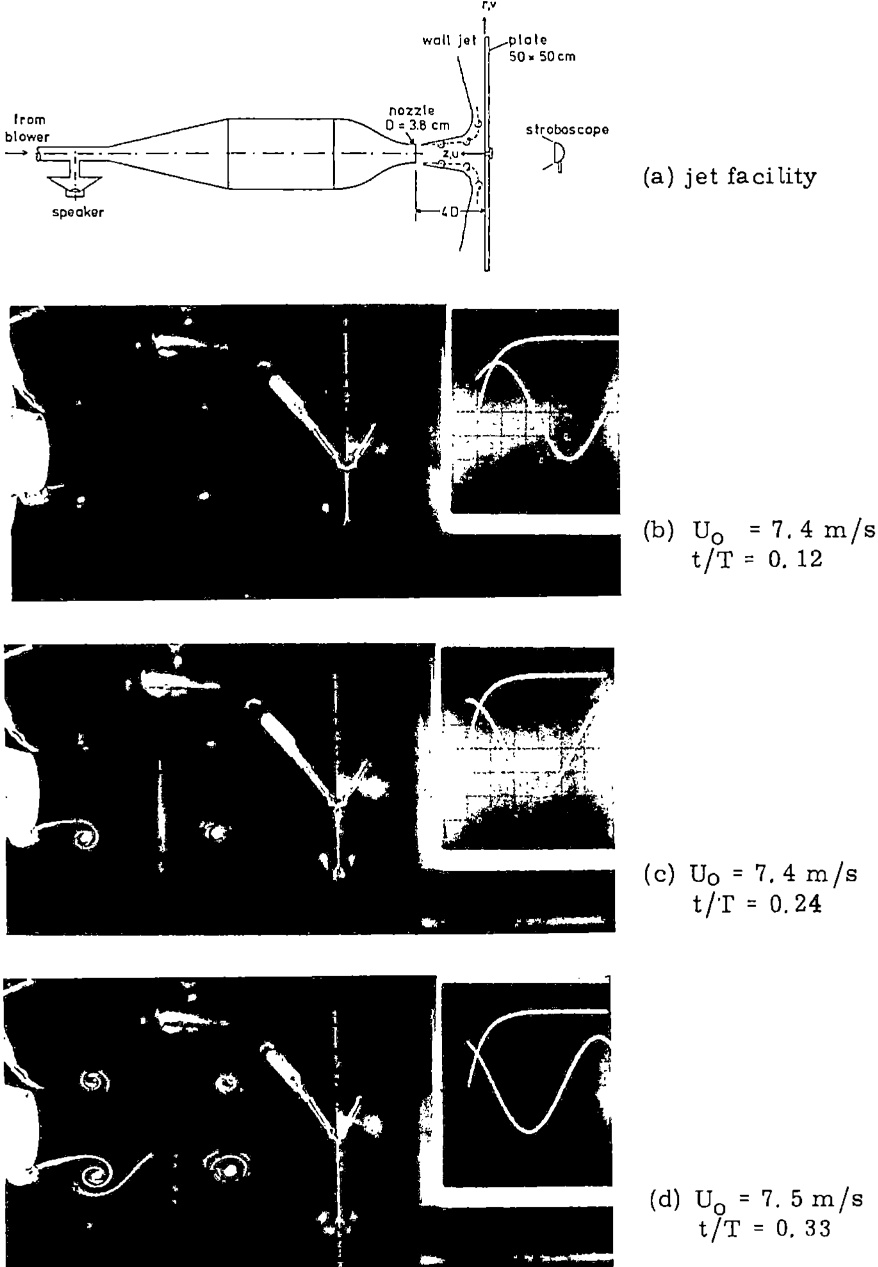

(a) jet facility

(b) U_O = 7.4 m/s
t/T = 0.12

(c) U_O = 7.4 m/s
t/T = 0.24

(d) U_O = 7.5 m/s
t/T = 0.33

Figure 5: (a): : Jet facility and impingement plate.
(b)-(d) : Smoke visualization in vertical symmetry plane of impinging jet. Stroboscopic illumination f = 70 Hz. Exposure time: 18 periods. View almost parallel to reflecting plate surface. Oscilloscope display: stroboscope trigger pulse and speaker signal of period $T = 1/f$.

number of 19 000. The loudspeaker upstream of the diffusor was driven by
a sinussoidal signal of forcing frequency f = 70 Hz, producing velocity
fluctuations at the jet exit with an amplitude u_{rms}/U_o = 0.18. The high
amplitude forcing was chosen both to achieve a highly coherent row of ring
vortices in the jet shear layer and to suppress pairing of ring vortices. The
jet impinged normally onto a plate 4 diameters downstream and spread
radially, forming an axisymmetric wall jet with ring vortices in the outer
shear layer. The cylinder-coordinate system r, z with corresponding
radial and normal velocities u, v is centered at the stagnation point.

For flow visualization two oil-smoke streaklines were produced in the
vertical symmetry plane in the jet shear layer (figure 5 b-c). Smoke of a
third streakline produced slightly below the stagnation point at the plate was
mainly entrained into the thin boundary layer and visualized the secondary
vortex formation. Illumination with stroboscopic light synchronized with
the speaker signal yielded a frozen pattern of the periodic flow. Exposure
over 18 periods illustrates that both the spatial jitter of the vortex path,
as well as phase fluctuations of the vortex passage at a fixed point in space
are reasonably small. This is particularly important for obtaining a good
representation of the flowfield of a single event by phase averaging the
periodic velocity signal. The zero phase of the speaker signal used as
reference is arbitrarily defined as that zero-crossing which has a positive
slope. In figure 5a the ring vortex is still approaching the wall, and the
boundary layer is attached. Shortly later (5b) the boundary layer has separated, and a secondary vortex moves downstream with the main vortex,
finally being wrapped around the main vortex (5c).

3.2 Experimental Procedure

Measurements of u-velocity profiles in a thin boundary layer (here
typically 1 mm) require a hotwire orientation parallel to the wall for sufficient spatial resolution. Near separation, however, the assumption $v \ll u$
is no longer valid and the hotwire measures the magnitude of the velocity
vector $q = (u^2 + v^2)^{1/2}$ normal to the wire. For measuring both velocity
components u and v simultaneously, a new hotwire sensor with two parallel
wires (PW-probe) shown in figure 6 was developed.[+] The first wire is operated at constant temperature (overheat ratio 20 %), and measures the
velocity q after being calibrated in the conventional manner. The second
' cold ' wire is operated at low constant current and is sensitive to temperature only. The second wire 70 µm downstream senses the temperature in
the laminar wake of the first wire, which only depends on the angle α and
the velocity q. The data of angle calibration at the jet nozzle for velocities
between 1 and 10 m/s and angles from $0°$ to $50°$ (at low velocities up to $80°$)
were fitted using a calibration function T (α, q). In the wall jet boundary
layer the PW-probe was oriented at an angle $\alpha_p = 30°$ to the plate. The
flow angle ß was determined by computing $\alpha(T, q)$ from the velocity q
measured with the first wire and from the output voltage T of the second
wire. The u- and v-signals were then computed from q(t) and ß(t). For data
acquisition and processing, a PDP 11/55 computer was used. The phase

---

[+] I should like to thank Dr. J. H. Haritonidis for suggesting this method
and for helpful advice.

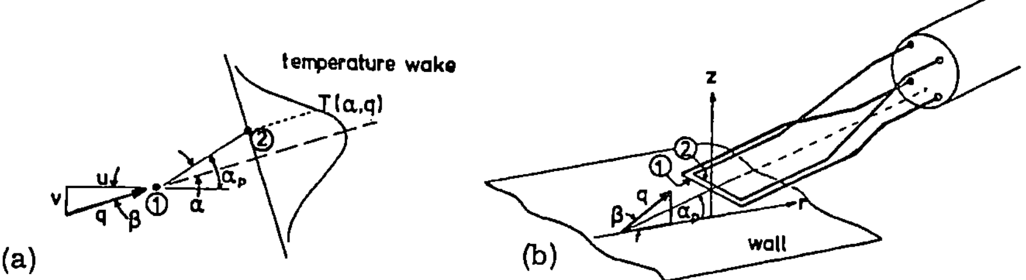

Figure 6 : Parallel-wire probe: (a) temperature wake of constant temperature wire 1 at location of wire 2. (b) sensor geometry and orientation.

averages $\hat{u}$ and $\hat{v}$ were computed chosing a characteristic feature of the velocity signal itself as sampling condition. The phase relation between different spatial points was provided by the speaker signal. For further details see Didden and Ho (1982). At all radial positions (spacing 0.1 D) the PW-probe was traversed normal to the plate in steps of 0.1 mm from z = 0.2 mm to 1.1 mm and in coarser steps at larger wall distance z. The wall distance was normalized with the time-averaged boundary layer displacement thickness z_1 = 0.25 mm at r/D = 0.9 (cf. figure 10).

3.3 Flowfield near Separation

The time-averaged velocity profiles $\overline{U}(z)$ of the wall jet are plotted in figure 7. In the impinging jet region between the stagnation point and the position r/D = 1.0 the maximum velocity at the edge of the boundary layer increases to 0.8 U_0. In the wall jet region at r/D $\geq$ 1.0 the boundary layer thickness increases, and the maximum velocity decreases. Starting at r/D = 1.1 a flattening of the velocity peak is observed, which is associated with boundary layer separation during part of each cycle, as will be seen from the phase averaged profiles. In the region with a flat profile (r/D = 1.5) the secondary vortex lifts off the plate and is wrapped around the primary vortex.

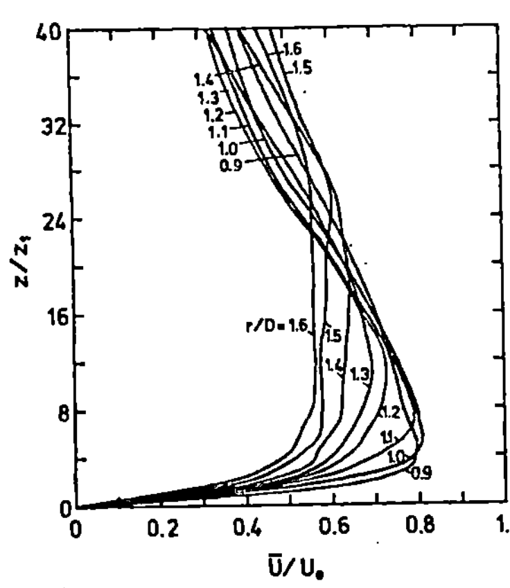

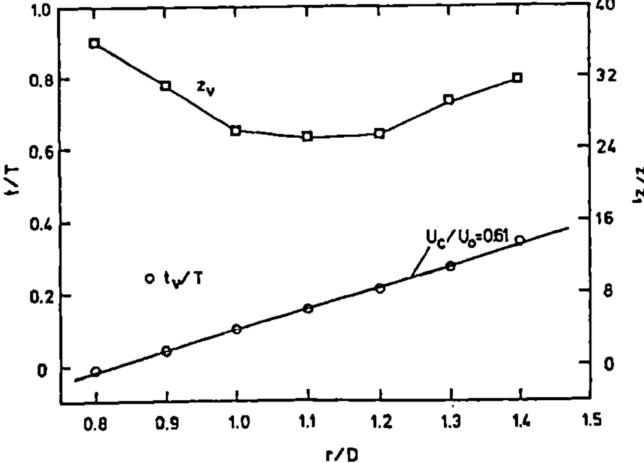

Figure 7. Time averaged velocity profiles $\overline{U}(z)$ of the wall jet.

Figure 8. Passage time t_V/T of the ring vortex and wall distance z_V of the vortex-axis versus r/D.

The passage time t_v/T of the ring vortex axis at each radial position was derived from the phase of the characteristic V-shaped minimum of the $\hat{q}$-signal through the vortex core. From figure 8 the convection velocity of the ring vortices travelling along the wall is seen to be approximately constant ($U_c = 0.61\, U_0$). The wall distance z_v of the vortex axis was determined by the z-position, at which the velocity $\hat{q}(z, t_v)$ at the passage time was equal to the convection velocity U_c. From $r/D = 1.0$ to 1.2 the ring vortices move almost parallel to the wall at a distance $z/z_1 \simeq 25$.

The phase averaged velocity profiles $\hat{u}(z)$ are plotted over one cycle in figure 9. The profiles are smoothed in z-direction in order to eliminate fluctuations in $\hat{u}(z)$ introduced by small errors in the phase relation of the signals $\hat{u}(t/T)$ at neighbouring z-positions. For all radial positions a pronounced peak is observed at $t \simeq t_v$ when the ring vortices in the outer flow pass the measuring point. The maximum velocity at the edge of the boundary layer is typically $\hat{u}_{max} = 1.6\, U_0$. At the radial position $r/D = 1.0$ a characteristic velocity defect and an inflexion-point profile in the boundary layer appear at $t < t_v$, i.e. downstream of the ring vortex. At the next position downstream at $r/D = 1.1$ the velocity defect increases and a region of negative shear velocity $\partial\hat{u}/\partial z$ distant from the wall is observed. In flows with steady boundary layer separation, the negative shear stress region downstream of the separation point is known to be attached to the wall. The appearance of $\partial\hat{u}/\partial z < 0$ detached from the wall is therefore characteristic of an unsteady separation moving downstream. The velocity profiles thus indicate that the separation occurs between $r/D = 1.0$ and 1.1 at approximately the phase $t/T = 0.1$. At $r/D = 1.2$ the irregular structure of the $\hat{u}$-velocity profiles near the wall at $t < t_v$ is most probably caused by the passage of the secondary vortex.

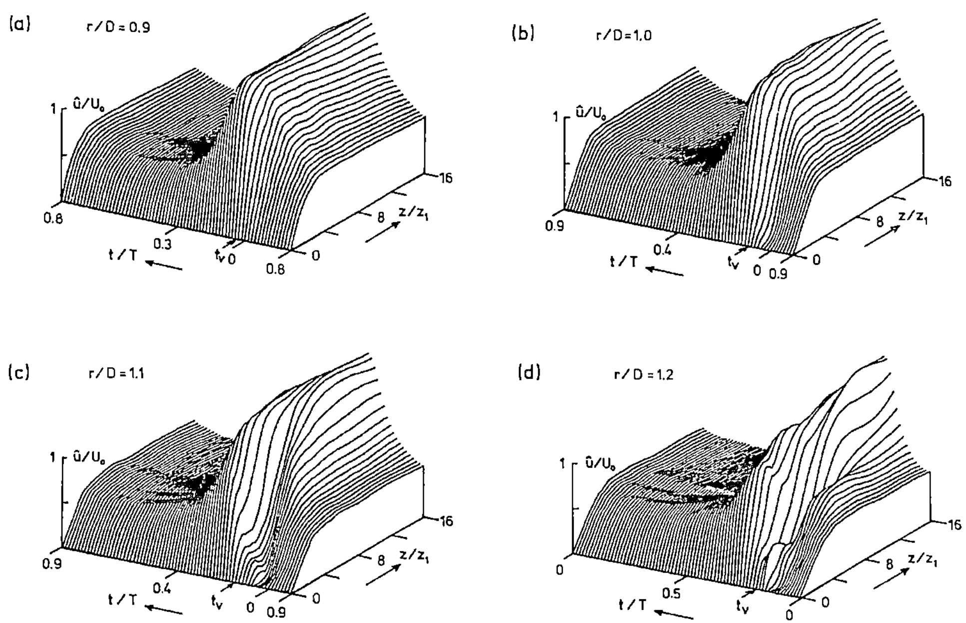

Figure 9. Phase averaged velocity profiles $\hat{u}(z)$ versus phase t/T.

The effect of separation and secondary vortex formation on the boundary layer is best illustrated by the change of the displacement thickness $\hat{\Theta}_d$ in figure 10:

$$\hat{\Theta}_d(t/T) = \int_0^{z_{max}(t/T)} [1 - \frac{\hat{u}(z, t/T)}{\hat{u}_{max}(t/T)}] \, dz, \qquad z_{max} = z(\hat{u}_{max}).$$

At $r/D = 0.9$ the variation of $\hat{\Theta}_d$ over one cycle is small, although the velocity $\hat{u}_{max}$, which defines the edge of the outer flow, varies between $0.45 \, U_o$ and $1.6 \, U_o$. Therefore the time average $z_1 = \overline{\Theta}_d$ at this position was chosen as a typical length scale of the undisturbed boundary layer. The minimum value of $\hat{\Theta}_d$ at each radial position is approximately equal to z_1.

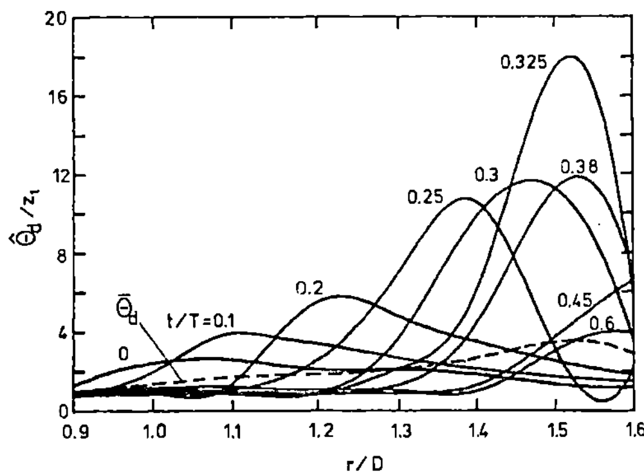

Figure 10. Temporal development of phase averaged displacement thickness $\hat{\Theta}_d$ versus r/D (interpolation with cubic spline function).

At each position the boundary layer thickening takes place over a short fraction of the period only. At $r/D = 1.1$, where the boundary layer is already separated, the displacement thickness varies by a factor 4 over each cycle. The maximum of $\hat{\Theta}_d$ corresponding to the maximum $\hat{u}$-velocity defect of figure 10 moves downstream ahead of the ring vortex and increases in magnitude as the secondary vortex increases in size. An absolute maximum of $\hat{\Theta}_d = 18 \, z_1$ is reached at $r/D = 1.5$. At this radial position, the secondary vortex is wrapped around the main vortex, resulting in a sudden reduction of the maximum displacement thickness further downstream at $r/D = 1.6$.

3.4 Vorticity Ejection

Further information about the flowfield at the initial phase of separation can be obtained from vorticity contours. Flow-visualization and velocity measurements indicate that the periodic flowstructure, in particular the separation region, moves downstream with a convection speed close to the convection velocity $U_c = 0.61 \, U_o$ of the ring vortex. Provided a vorticity structure of streamwise length scale l in the wall jet is steady over a timescale l/U_c in a frame of reference moving downstream at velocity U_c, the vorticity distribution of this structure is

$$\hat{\omega}(z,\ t/T) = \frac{\partial \hat{u}(z,\ t/T)}{\partial z} + \frac{1}{U_c}\ \frac{\partial \hat{v}(z,\ t/T)}{\partial t}$$

and locally, over a lengthscale l, the time coordinate can be replaced by the spatial coordinate $-r/U_c$.

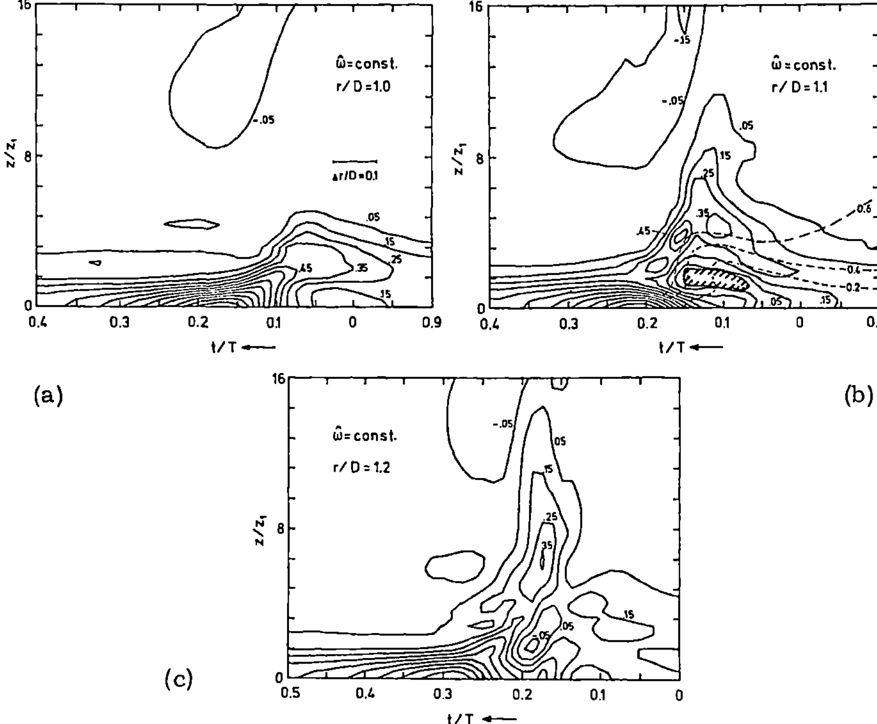

Figure 11. Vorticity $\hat{\omega}/\omega_1$ = const. versus phase $-t/T$ over half a period $(\omega_1 = U_0/z_1)$. Spatial distance $\Delta r/D = 0.1$ corresponds to $\Delta t/T = -0.058$. The shaded area indicates shear stress $\tau = 0$, — — — $\hat{u}/U_0$ = const.

With these restrictions in mind the vorticity contours $\hat{\omega}$ = const of figure 11 plotted against $-t/T$ over half a period, give some indication of the spatial vorticity distribution, with radial direction $r = -U_c t$ from left to right. The z-scale of the contour plot is enlarged by a factor 6.2 relative to the radial scale. Comparison of the plots at $r/D = 1.0$ and $r/D = 1.1$ gives an indication for the deformation of the flow pattern as it moves 0.1 D downstream, which in the time domain corresponds to $\Delta t/T = 0.058$ (marked in figure 11a). At all radial positions the vorticity maximum at the wall is due to the high shear induced by the passing ring vortex. At $r/D = 1.0$ boundary layer vorticity starts to lift off the wall. Further downstream at $r/D = 1.1$ a sheet of vorticity is ejected into the outer flow. It is at this position that for the first time a local vorticity maximum separated from the boundary layer is observed at a wall distance $z/z_1 = 4$. Further downstream at $r/D = 1.2$ the vorticity maximum has moved to a larger wall-distance. In figure 11 b the region of negative shear stress $\tau = \mu(\partial \hat{u}/\partial z + \partial \hat{v}/\partial r) \leq 0$ at $r/D = 1.1$ is marked by the shaded area. The significance of shear stress $\tau = 0$ for separation will be discussed in the next chapter.

3.5 Discussion and Conclusion

Based on two-component velocity measurements with the 'parallel-wire'-sensor, the phase averaged $\hat{u}$-velocity profiles, the boundary layer displacement thickness and the vorticity distribution were computed at radial positions close to the onset of the unsteady, downstream-moving separation.

In the following we discuss the features characterizing the vortex-induced separation. Boundary layer separation, in general terms, is characterized by the 'break away' of boundary layer fluid from the wall, resulting in a change of the outer flow. For steady flow with adverse pressure gradient the separation is well known to occur at a point of vanishing wall shear stress $\tau_w = \mu(\frac{\partial u}{\partial z})_w = 0$. In the case of unsteady separation considered here, the initial location of separation is most clearly indicated by the vorticity ejection into the outer flow. Although, due to the unsteadiness of the flow the vorticity contours in figure 11 are an approximate representation of the spatial vorticity distribution only, the contour plots show that vorticity ejection, and thus separation, starts at a location between $r/D = 1.0$ and 1.1. This vorticity ejection results in a change of the outer flow, which becomes apparent at $r/D > 1.2$ as the ring vortex is driven away from the wall (figure 7). The initial location of separation is further confirmed by the plot of phase averaged $\hat{u}$-velocity (figure 9) with inflexion-point profiles at $r/D = 1.0$ and profiles with negativ shear velocity $\frac{\partial \hat{u}}{\partial z} < 0$ at $r/D = 1.1$. Another feature, which is also typical for steady separation, is observed at $r/D = 1.1$ (figure 11 b): the region of negative shear stress (shaded area) is located between the wall and the sheet of displaced vorticity, which originates from the attached boundary layer upstream. Different from the case of steady separation, however, is the fact that the region of negative shear stress is detached from the wall, and that no reverse flow at the wall exists.

On the basis of theoretical considerations several investigators recognized that the separation condition $\tau_w = 0$ with reversed flow at the wall downstream of separation, is not significant for unsteady separation. Moore, Rott and Sears proposed a separation condition for unsteady flow, which became known as the MRS-condition (see for example the review article by Sears and Telionis, 1975). These authors postulated that separation occurs at that point within the boundary layer at which the shear stress and the velocity vanish in a frame of reference moving with separation, i.e. the separation point is defined by $\tau = 0$ at $\hat{u} = U_s$. We dont know of any comparison of the MRS-condition with experimental results for downstream-moving separation - in fact, velocity measurements seem to be lacking for this case altogether. According to our measurements the vortex-induced separation region is known to move downstream at approximately the convection velocity of the ring vortex, i.e. $U_s \simeq U_c = 0.61 \, U_o$. The zero shear stress points in figure 11 b, however, are located in a region with local velocity $\hat{u} \leq 0.4 \, U_o$, which is much smaller than the velocity of the separation region. The MRS-condition therefore is not satisfied for the vortex-induced separation. This result is consistent with boundary layer calculations by Walker (1978) of the separation induced by a straight vortex moving parallel to a flat plate. Walker also found zero shear stress only in regions with a velocity smaller than the separation velocity.

Since separation is expected to occur in flows with an adverse pressure gradient only, with the following arguments we will provide evidence that the separation observed here is caused by the pressure field of the vortex moving along the plate. In absence of the impinging jet a single vortex produces an adverse pressure gradient $(\partial p/\partial r)_v > 0$ due to the diverging flow ($v > 0$) in the wall region downstream of the vortex. In the stagnation flow of the impinging jet a favourable mean pressure gradient $\partial \bar{p}/\partial r < 0$ exists at the wall. At $r/D = 1.0$, the location at which the mean velocity $\bar{U}$ (figure 7) at the edge of the boundary layer reaches a maximum ($\partial \bar{U}/\partial r = 0$), the mean pressure gradient becomes small. Therefore, at some radial position close to $r/D = 1.0$, one would expect the magnitude of the favourable mean pressure gradient $\partial \bar{p}/\partial r$ to become smaller than the pressure gradient fluctuations due to the vortex flow. Then, in the wall region downstream of the vortex, the pressure gradient becomes positive (adverse pressure gradient) - a prerequisite for separation. The vorticity ejection at $r/D = 1.1$ confirms that the above situation indeed exists at a radial position $r/D < 1.1$.

The secondary vortex formation is therefore initiated by the adverse pressure gradient of the main vortex, which causes fluid of the boundary layer to be displaced from the wall. The distribution of ejected vorticity is deformed by the vortex flow field, and by the self-induced flow of the ejected vorticity, tending to roll up the free end of the sheet-like vorticity distribution into a secondary vortex. Since the region of adverse pressure gradient moves downstream with the ring vortex, the secondary vortex is continuously supplied with boundary layer vorticity. It thus increases in size and strength until it is lifted off the wall under the influence of the main vortex.

Acknowledgement

The measurements on unsteady separation were performed at the Department of Aerospace Engineering, USC. I wish to thank Prof. J. Laufer, Chairman, and Prof. Ho Chih-Ming for giving me the opportunity to work in such stimulating surroundings. Essential support and helpful discussions with Prof. Ho have greatly contributed to this study.

References

Blendermann, W. (1969) Der Spiralwirbel am translatorisch bewegten Kreisbogenprofil; Struktur, Bewegung und Reaktion. Schiffstechnik 16, 3-14

Cerra, A. W. and Smith, C. R. (1980) Experimental Observation of the Interaction of a Vortex Ring with a Flat Plate. Bull. Am. Phys. Soc. 25, p. 1092

Didden, N. (1977) Untersuchung laminarer, instabiler Ringwirbel mittels Laser-Doppler-Anemometrie. Mitt. aus dem MPI für Strömungsforschung und der AVA Nr. 64, Göttingen

Didden, N. (1979) On the Formation of Vortex Rings: Rolling-Up and Production of Circulation. Journal of Appl. Math. and Phys. (ZAMP), 30, 101 - 116

Didden, N. and Ho, Chih-Ming (1982) Unsteady Separation in the Boundary Layer of an Impinging Jet. Manuscript in preparation

Harvey, J. K. and Perry, F. J. (1971) Flowfield produced by trailing vortices in the vicinity of the ground. A. I. A. A. Journal 9, 1659 - 1660

Kaden, H. (1931) Aufwicklung einer unstabilen Unstetigkeitsfläche. Ing. Arch. 2, 140 - 168

Liess, C. and Didden, N. (1976) Experimente zum Einfluß der Anfangsbedingungen auf die Instabilität von Ringwirbeln. Z. Angew. Math. Mech. 56, T 206 - T 208

Maxworthy, T (1977) Some Experimental Studies of Vortex Rings. J. Fluid Mech. 81, 465 - 495

Pullin, D. I. (1978) The large-scale structure of unsteady self-similar rolled-up vortex sheets. J. Fluid Mech. 88, 401 - 430

Pullin, D. I. (1979) Vortex ring formation at tube and orifice openings. Phys. Fluids 22, 401 - 403

Saffmann, P. G. (1978) The Number of Waves on Unstable Vortex Rings. J. Fluid Mech. 84, 625 - 639

Schneider, P. E. M. (1980) Sekundärwirbelbildung bei Ringwirbeln und in Freistrahlen. Z. Flugwiss. Weltraumforsch. 4, 307 - 318

Sears, W R. and Telionis, D. P. (1975) Boundary Layer Separation in Unsteady Flow. SIAM Journal on Applied Mathematics, 28, 215 - 235

Wedemeyer, E. (1961) Ausbildung eines Wirbelpaares an den Kanten einer Platte. Ing. Arch. 30, 187 - 200.

Walker, J. D. A. (1978) The boundary layer due to rectilinear vortex. Proc. R. Soc. Lond. A 359, 167-188.

VORTICES FOLLOWING TWO DIMENSIONAL SEPARATION

by

A. Dyment

Université de Lille I
Institut de Mécanique des Fluides
Lille, France

<u>Table of principal symbols</u>.

b	vortex spreading rate parameter
h	vortex street spacing
H	thickness of the body
k	dimensionless propagation velocity V/U
L	characteristic length of the body
M	Mach number
N	vortex frequency
r	vortex radius
R	Reynolds number UL/ν
S	Strouhal number NL/U
t	time
U	characteristic velocity
V	propagation velocity of vortices
W	propagation velocity of the vortex street structures
x	curvilinear abscissa on vortex trajectory
Γ	circulation
Δt	time interval between two successive flashes
λ	vortex street wave length
ν	kinematic viscosity
ξ	dimensionless abscissa x/L

Subcripts : a   vortex street
             e   vortex shedding
             o   end of recirculating domains.

<u>Introduction</u>.

       This paper is devoted to the ordered properties of vortices which
follow a two dimensional separation in a high Reynolds number flow around a
fixed body. Such phenomena are quite familiar, but in fact they are very
complicated and badly known. The most surprising property is their
unsteadiness although boundary conditions do not depend on time. An important
literature deals with this problem. The aim of our communication is not to
do an extensive bibliographical review : only the main results will be
pointed out, including some new informations issuing from our own work.

       In the light of the experiments, it seems that any attempt to
model the near wake in a steady state motion framework is hopeless. However,
the high Reynolds number hypothesis leads to some simplification because
the instability zone of any shear layer following separation has very
small dimensions. Thus, the vortices can be considered as shed from the

separation point. The unsteady phenomena induced by this emission can be taken into account with a simplified scheme. That scheme allows the successive vortex pairings in the mixing layer to be explained and the conditions of formation of vortex streets clarified. The proposed laws, expressed in terms of order of magnitude can obviously provide only a rough sketch of the complex phenomena occurring in the near wake. Their main weakness is to split the flow into different domains in which only the dominant features are taken into account and therefore possible couplings are ignored. In fact, such couplings exist and experiments show that they can be intense, especially in supercritical flow.

1. Separation.

The first theory dealing with separated steady flows over two dimensional bluff bodies has been worked out for an incompressible inviscid fluid by Helmholtz and developed by Kirchoff (see for instance [1]). The proposed model admits a stagnant fluid wake extending to infinity in order to obtain a non zero drag. As a result, the pressure inside the wake is equal to the upstream free flow pressure. That theory has several failures. At first, the drag coefficient which it gives is to small. Furthermore, the measurements show that the base pressure coefficient is not zero. Finally, an unbounded dead fluid wake is not physically acceptable because the disturbances created by the body must vanish at infinity.

A possible improvement is to admit this theory as a first order approximation for high Reynolds number flows : this leads to adding boundary layers to the walls and the separation lines. But then another failure appears : since near the separation point the streamline issuing from it has a finite curvature, the upstream pressure gradient is favourable, so that there is no reason for the boundary layer to separate.

Prandtl and later Batchelor [2] have proposed a different model. They assume the existence of recirculating domains of dimensions comparable to L . This yields that inside the recirculating domains the velocity is of the same order of magnitude as U [3] . It is possible to prove that, if $R \to \infty$, the recirculating flow is governed by Euler equations with a constant vorticity (but this constant remains unknown).

From a purely phenomenological point of view the Prandtl-Batchelor model seams more realistic than Kirchoff's. However, some reserves must be raised. First, measurements show that the recirculating flow velocity is clearly smaller than U . Furthermore, due to the finite dimensions of the recirculating domains, the drag goes to zero when $R \to \infty$. Batchelor states that if the Kirchoff's model is suitable for an inviscid fluid $(R = \infty)$, it cannot be applied for $R \to \infty$. In other words, it would be a singular perturbation problem. A good argument for this statement is the following : there is no reason for the separation point from Kirchoff theory to coincide with the zero tangential stress point when $R \to \infty$.

In 1967 Sychev [4] tried to solve these difficulties by assuming a drag coefficient of order unity and bounded recirculating domains. These hypothesis led him to consider a new model in which the recirculating domains have LR as characteristic length and $LR^{1/2}$ as characteristic width whereas the thickness of the wake is of order L . The fluid is almost stagnant inside the recirculating domains which then have an elliptic form. Sychev distinguishes different domains which are matched togather in the asymptotic development sense. The drag corresponds to Kirchoff's theory one, the first corrective term being of order $R^{-1/2}$.

Later, Sychev used the triple deck structure to solve the separation problem for incompressible flows [5] . That structure implies the existence of a weak pressure gradient singularity (of order $R^{1/8}$) which acts on a short distance (of order $R^{-3/8}$). That result disagrees with the usual belief that separation is caused by an adverse pressure gradient being applied on a finite length. Sychev found that for $R \to \infty$ the separation vicinity is correctly described by Krichoff's model. Smith [6] improved the previous model. He achieved the calculations for the circular cylinder. The comparison with experiments for $R < 10^2$ seems to confirm the validity of this model, but such comparison is questionable since the small parameter $R^{-1/8}$ entering into the theory is equal to .56 for $R = 10^2$!

Of course, for higher Reynolds number experimental check is no possible due to the instability of the flow. Moreover, for $R \gg 1$, the recirculating domains have dimensions of order L and from this point of view the Prandtl-Batchelor model is more suitable. Finally, the paradox of the asymptotic Sychev-Smith model is that it seems appropriate for $R \sim 10^2$ but not for higher values of R because of the unstable character of the Navier Stockes equations solutions.

Therefore, the steady state schemes of separation and wake are not able to give a correct representation of the phenomena. That means that at high Reynolds numbers the unsteady effects are not simply superimposed to a mean steady motion, constituting somehow a perturbation field, but that their coupling with the mean motion must not be ignored from the first approximation study. This is not to surprise the experimentalists who have noticed for a long time that separated flows present unsteady properties characterized by the existence of large scale eddies for which the treatment by classical turbulence methods is very questionable.

The conditions for the appearance of these eddies, their streamwise evolution, their degeneration into fine scale turbulence make as many unknowns in spite of the intense effort which is presently made in this field.

2. Mixing layers.

The streamline separating the flow coming from upstream and the recirculating flow belongs to a shear layer. Shear layers are often met and numerous studies have already been devoted to them (see for instance [7]). From a theoretical point of view, the simplest problem concerns the stability of an infinite shear layer, initially rectilinear. This flow does not constitute an exact picture of a real shear layer following separation ; nevertheless, the results which have been obtained are significant of a major feature : the evolution towards the birth of large eddy structures. In the present case this tendency is timewise ; for a real mixing layer the spatial character is dominant.

This property, forecasted by calculations, has been pointed out experimentally, in particular by Winant and Browand [8] and by Brown and

Roshko [9] . Different works on the same topic have been achieved later, for which a short review can be found in [10] .

It is usual to call the flow under consideration a turbulent mixing layer. This is surprising for the flow is not, at least at the beginning, chaotic with random motions in a wide range of frequencies. Our approach is quite different : we take into account the existence in a steady separated flow of a small domain which surround the separation point and where the flow is governed by the full Navier Stokes equations [3] [11] . Outside this domain the shear layer is unstable since its representative Reynolds number is large. This instability gives birth to vortices which are formed over a very short distance with respect to L . Thus, it is natural to admit that the vortex shedding is a local property depending only on characteristic quantities of the Navier-Stokes domain. Furthermore, if viscosity does not interfere directly, we obtain [12] $S_e \sim R^{2q-1}$, $\Gamma_e \sim \nu$ with $q > 1/2$. The formed vortices are very small and very concentrated, so that their diffusion is taken into account. Meanwhile, the convection effects are dominant which means $V \gg dr/dt$, with $b^2 d(\tau^2) = \nu dt$. The spreading rate b and the dimensionless velocity $k = V/U$ depend only on $\xi = x/L$.

It has been proven [10] [12] that the preponderance of convective effects yields the result that two successive vortices necessarily meet and agglomerate together. The abscissa of the beginning of the pairing process (subcript 1) is given in terms of order of magnitude by $\xi_1 \sim b_1^2 k_1^3 R S_e^{-2}$.

The detected overlapping phenomenon will repeat, so that a succession of pairings takes place. After the n th pairing we get $S_n = 2^{-n} S_e$, $\xi_n \sim 4^{n-1} \xi_1$, $\Gamma_n \sim 2^n \nu$. These formula show that the successive pairing process slows down quickly as we progress downstream [10] [12] [26] .

It is convenient to represent the successive pairings by a continuous evolution. Thus it comes $[10] R k^2 = \alpha S^2 \int_0^\xi \frac{d\xi}{k b^2}$ or $V^2 = \alpha \nu N^2 \int_0^\xi \frac{dx}{V b^2}$, with $4 \leqslant \alpha \leqslant 16$. From the order of magnitude view point this yields $S^2 \xi \sim b^2 k^3 R$. This relationship is relatively well verified, even for high speed flows (fig. 1).

Is our approach reconcilable with that considering the vortex pairing as an essentially random phenomenon ? Although these points of view look totally opposed, it must be kept in mind that in our scheme the velocity field induced by the travelling vortices has been neglected. Moreover, a splitting between diffusion and convection processes has been implicitely admitted. Such approximations are probally valid at the beginning of the mixing layer. On the other hand, in the flat part of the curve giving S versus ξ (fig. 1), where the frequency becomes approximately constant, the phenomenon seems to loose most of its ordered character as can be seen from the visualization : in this zone, instead of merging, vortices rather gather in clusters [10] .

Finally, the set of analyzed phenomena can be summarized as follows : downstream of the Navier-Stokes domain and the region of formation of vortices which both are almost reduced to a point in comparison with the body size, two zones can be distinguished. . A first one with well ordered pairings induced by diffusion, then a second one where the pairing process

is pseudo random. The large eddy structures do not persist to infinity.
Their spontaneous destruction and their degeneration into fine scale
turbulence is most probably a three dimensional phenomenon.

3. Vortex streets.

Fluid mechanicists have always been puzzled by the existence of
vortex streets in the wake of two dimensional bodies. As it has been
previously seen, the unsteady character of a flow with fixed boundaries
can be explained by separation induced instability. But, it may look
surprising that a symmetric body flow displays such a strong asymmetry.
However, visualizations show well that the large vortex structures overlap
to form an asymmetric configuration which persists far downstream. That
can be justified as follows : the two separations which are generally
found on the body produce two rows of vortices with circulations of
opposite sign and same order of magnitude. Thus, neither can get the
advantage over the other. The result is that the vortices can interact
at the end of the recirculating domains and give rise to a vortex street.
The reason of its long existence can be explained if we suppose that each
eddy propagates with the same velocity W . It is clear that, in the frame
of reference moving with the velocity W (fig. 2), at the border of two
neighbouring cells the fluid moves in the same direction so that there is
no tendency towards destruction.

The classical Karman theory provides a model in inviscid flow
consisting in two parallel rows of point vortices [13] . Karman showed
that there is only one possible vortex configuration corresponding to
$sh\,\pi h/\lambda = 1$, i.e. $h/\lambda = .28$. In fact, it has later been proven that this
relationship constitutes only a minimum instability condition [14] .

Although the viscosity plays a minor role in this phenomenon as
the vortex street disposition is almost independant of R , it must be
noticed that the theory has been extended to viscous fluids in the Oseen
approximation framework [15] .

With the new possibilities provided by modern computers, numerous
numerical simulations have been performed in order to reproduce the vortex
streets. Some of them [16] start with two parallel infinite rows of point
vortices to which a perturbation of wave length λ is applied. The
computation reproduces the time evolution of the flow and shows that the
vortices gather into alternate packs : .28 is the minimum value of the
ratio h/λ for which only two packs form on one wave length.

Others [17] have undertaken a similar calculation but with
pseudo-viscous vortices, e.g.comprising a rotating core. The main result
of that is the existence for the ratio h/λ of a core dimension dependant
interval around the .28 value corresponding to a stable street configuration. This would mean that the viscosity has a stabilizing effect.
Therefore, the Karman condition would be valid as soon as viscosity, even
small, exists.

All these calculation attempts give results the application of
which must be restricted to the far wake since the presence of the body is
absolutely ignored.

When the body is to be taken into account two ways are possible. The Navier-Stokes equations can be solved directly ([18] for instance). But it is well known that the larger the Reynolds number, the finer the calculation grid must be, so that the method is limited to values of R of about 10^3 . Another method lies in simulating the phenomenon in inviscid fluid by a vortex injection near the separation point. The free vortices injection parameters are adjusted to match the experimental results. The intensity of the introduced vortices is related to the velocity outside the boundary layer, at separation. The vortex injection point is determined by the Kutta-Joukowsky condition when the separation occurs at an angular point, but if the separation point is not a priori known a new difficulty arises. The injection is in general made at a distance from the body which is equal to the boundary layer displacement thickness and its frequency remains arbitrary. Interesting results have been obtained [19] [20] , but the method can only constitute a model since it ignores the viscosity effects at separation. Furtherwore, the results do not agree completely with experiments.

Concerning the knowledge of the phenomenon, the cylinder is certainly the most studied case, though the corresponding flow is not among the simplest ones because the separation location is not fixed. Different regimes are distinguished in literature : for $R \leqslant 300$ the frequency N_α depends on R ; for $300 \leqslant R \leqslant R_1$ the separation is laminar and the Strouhal number S_α is approximately constant ; for $R_1 \leqslant R \leqslant R_2$ there is a separation bubble followed by an irregular wake ; for $R \geqslant R_2$ the separation is turbulent and a certain periodicity appears again in the wake. The separation bubble regime is not well known and the authors do not agree for the values of R_1 and R_2 . According to some authors S_α goes suddenly from .20 to .45 for $R \simeq 4 \; 10^5$ but this result is not corroborated by others.

The experiments do not provide any explanation for the vortex street formation the mechanism of which remains uncleared. The only positive information is that the vortex street results from the interaction of the two vortex rows created at the separation points. To check that, it is sufficient to fit the body with a splitter plate. The plate separates the two mixing layers without noticeably disturbing the vortex shedding because S_e depends only on the local properties, and it prevents the contact between the two vortex rows : therefore, the vortex street formation is delayed and can be completely suppressed provided that the plate be long enough.

The use of our evolution scheme for the mixing layer vortices is able to clarify some points. Let us consider a symmetric flow (fig.3). Since L_o is of order L it is shown [12] that S_o is large whereas all experimental results give S_α of order one. Consequently the vortices of each mixing layer must undergo a new coalescence, completely different in nature from those considered in section 2. That is done quickly if S_o is not far from S_α , that is if R is moderately large : in this case the vortex street appears immediately behind the body. On the other hand, if R is very large, S_α is much smaller than S_o and the coalescence requires a long time : then, the vortex street is formed far downstream. The previous results can also be found by introducing a global relaxation time necessary for the vortex frequency to change from N_e to N_α [26] .

For a thin body, L_o is of order H and it is possible that the two mixing layers interact before the end of the recirculating domains. In spite of that, vortex streets are observed in airfoil wakes. For instance, on figure 4 which represents a supercritical flow with a separation bubble,we can distinguish a vortex street forming about half a chord length downstream of the trailing edge. Thus, such a street appears even when two separation points exist on the same side of the body. A similar phenomenon probably occurs on the cylinder within the separation bubble regime : the street formation is delayed and it may not be detected if the observation field is too reduced.

The coalescence of the mixing layer vortices to form the vortex street structures is obvious on figures 5 and 6 : at the beginning each vortex street cell is made of a cluster of vortices in the process of merging. Then, after some length, each cell appears as made up of a single vortex.

It has been noticed [21] that, in case of turbulent separation, the mixing layer vortices have a smaller frequency. Then the transition from S_o to S_a is easier and the vortex street can be formed closer to the body than with a laminar separation at high Reynolds number. Experiments show that the separation nature has no influence on the vortex street.

It may seem surprising that the mixing layer vortices tend to give birth to a vortex street, since this requires a sudden frequency adjustement. The reason probably lies in the fact that the vortex street arrangement leads to the minimum energy dissipation within the wake.

4. Coupling and compressibility effects.

Is it allowable, at least as a first approximation, to suppose that the flow around the body is independant of the unsteady motions produced in the wake ? That was the model presented by Roshko [22] who has limited the coupling to an intermediate zone between the near wake, assumed to be almost steady, and the vortex street situated in the far wake.

That model seems insufficient. Hot wire measurements close to the cylinder wall have shown [23] that the separation undergoes oscillations induced by the vortex street. From that, we must wonder whether our phenomenological theory for the vortex shedding remains valid, because it has been elaborated on a steady separation scheme. The answer is positive for S_e is very large compared to S_a . However, the pairing process is certainly disturbed and gets a more random character.

We shall now see that compressibility will increase the coupling. On this point, a large part will be made to our own laboratory experiments since informations about the unsteady properties of high speed flows are very rare in the literature. These experiments use essentially a Cranz-Schardin high speed visualization system [24] which allows to restore the time history of the observed flow during a given interval [10] [21] [25] [26] [27] [28] . Figures 1,5,6 and 7 show some examples of such visualizations.

There is a priori no physical reasons why compressibility should modify in a sensible way the shedding of vortices, their propagation mode and their pairing process. Therefore, it may be expected that the incompressible flow results be extended, at least qualitatively, to high speed flows. Such statement looks reasonable since some agreement has been found between the theory established for incompressible flow and some subsonic experiments (fig.1).

The coupling process becomes more and more intense with increasing Mach number. It appears as a periodic oscillation of the two mixing layers and a periodic motion of the separation when it occurs at a finite curvature point. It has been verified that the frequency of these fluctuations is equal to N_α . They are less intense in case of turbulent separation : it seems that the wake oscillation is impeded by the small scale motions.

An important effect of compressibility is the appearance of acoustic waves. For this point, vortices appear as sound sources. Wavelets can be seen on the pictures near each vortex kernel, especially when the local velocity is close to the speed of sound. But the strong waves which go upstream around the body are shed by the vortex street structures as the analysis of visualization pictures shows that their frequency is close to N_α . That can be checked when the waves are quasi normal to the flow direction (fig. 6) because the distance between two successive waves is about $(a'-U')/N_\alpha$, U' and a' being respectively the fluid velocity and the speed of sound above the body [25] [26] .

When the flow becomes supercritical, new phenomena appear because of the presence of shock waves. The existence of an unsteady field superimposed on a steady mean motion suggests the introduction of the notion of intermittently supercritical flow : it is a flow where supersonic domains only appear transitorily. On the other hand, the flow can be called strictly supercritical if the velocity remains always supersonic in a domain however small it might be. In such flows shock waves exist permanently and an interaction must be expected between these waves and the vortices shed at separation, which is generally located upstream. This is the case when the interaction between a shock wave and a boundary layer induces a separation a little upstream of the shock. This is also the case for a bluff body when supersonic pockets exist close to the recirculating domains where the fluid undergoes a strong expansion.

The coupling between the vortex street and a cylinder has been described previously [27] . The whole of our observations, briefly summarized on figure 8, can be explained by arguing on the possibility of the upstream transmission of the disturbances issued in the vortex street. As soon as the flow becomes strictly supercritical a shock wave exists and it prevents the coupling between the vortex street and the near wake. This means that the coupling must be maximum for the critical value of M , and indeed this corresponds to observations.

When the size of the supersonic pockets is bounded the upstream influence can take place both by going around the top of the shock wave and through the recirculating domains (fig. 8, M = .70). At a larger Mach number the shock waves extend far out and the disturbances can only be transmitted upstream through the subsonic part of the recirculating domains, which become smaller as M increases (fig. 8, M = .90). Above a given

value of M   there are two λ shock waves the feet of which almost come
into contact so that the communication between the vortex street and the
near wake is cut off : in this manner the whole domain situated upstream
of the shock waves becomes independant of the vortex street, that is to
say quasi steady (fig. 8, M = .98). The existence of this steady regime
yields an explanation of the sudden increase of S_a which has been noted
when M approaches 1 (fig. 9) : as soon as the steady regime is reached
the size of the body no more represents the characteristic length of the
vortex street. If me consider that length to be now the spacing between the
two shock waves, and if a Strouhal number is formed with this distance, we
find again a value close to .2.

More detailed quantitative results have been obtained by
association of the high speed visualization with unsteady pressure
measurements [27] [28] . An example is given on figure 7 which shows the
Fourier spectrum of the transducer signal at azimuth 85° and the same signal
after filtering. The pressure fluctuation level is comparable to the mean
pressure. The pressure integration gives an important unsteady lift force :
at M = .75 the unsteady lift component is about 15 % of the mean
drag [28] .

<u>References</u>.

1	Wu	An. Rev. Fl. Mech. 4, 243
2	Batchelor	J. Fl. Mech. 1, 388
3	François	Publ. ONERA n° 128
4	Sychev	Symp. Mech. Fl. and Gases, Tarda, 1967
5	Sychev	Mek. Zhid. i Gaza, 3, 47
6	Smith	J. Fl. Mech. 92, 171
7	Betchov, Criminale	"Stability of Parallel Flows", Pergamon, 1967
8	Winant, Browand	J. Fl. Mech. 63, 237
9	Brown, Roshko	J. Fl. Mech. 64, 775
10	Dyment	"Unsteady Turbulent Shear Flows", 359, Springer, 1981
11	Stewartson	Adv. Appl. Mech. 14, 145
12	Dyment	Comptes Rendus Ac. Sci. 290, B, 47
13	Villat	"Leçons sur la théorie des tourbillons" Gauthier Villars, 1930
14	Kochin, Kibel, Roze	"Theoretical Hydromechanics" Interscience, 1964
15	Wille	Adv. Appl. Mech. 6, 273
16	Abernathy, Kronauer	J. Fl. Mech. 13, 1
17	Christiansen, Zabusky	J. Fl. Mech. 61, 219
18	Fromm, Harlow	Ph. Fluids 6, 975
19	Sarpkaya	J. Fl. Mech. 68, 109
20	Clements, Maull	Prog. Aero. Sci. 16, 129
21	Dyment, Gryson	Colloque AAAF, Marseille, 1978
22	Roshko	J. Aero. Sci. 22, 124
23	Dwyer, Mc Croskey	J. Fl. Mech. 61, 753
24	Merzkirch	"Flow Visualization" Academic Press, 1974
25	Dyment,Gryson,Flodrops	2nd Symp. Flow Visualization, Bochum, 1980
26	Dyment, Gryson	AGARD CP n° 227, 1978
27	Dyment,Gryson,Ducruet	Euromech Coll. n° 137, Marseille, 1980
28	Rodriguez	Euromech Coll. n° 160, Berlin, 1982.

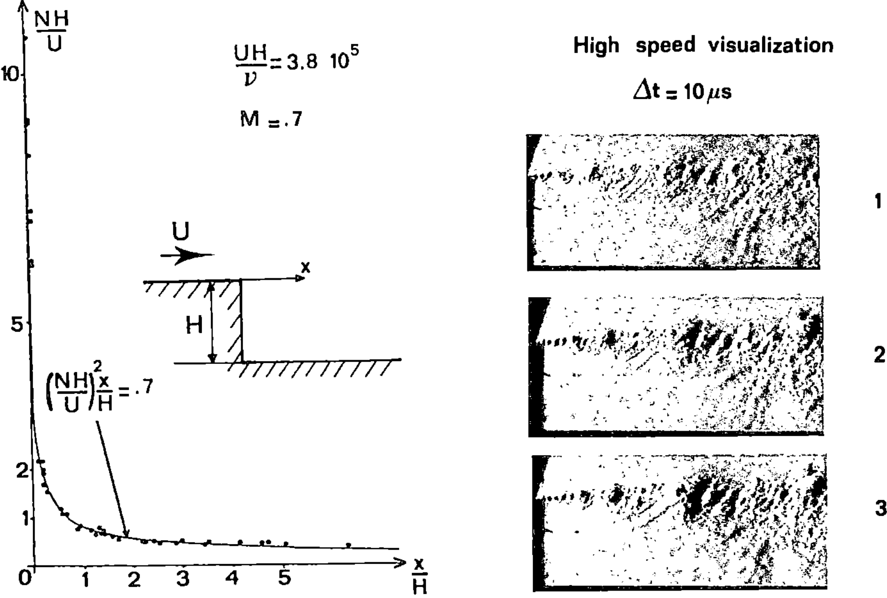

Fig:1   The mixing layer following a backward facing step   [10]

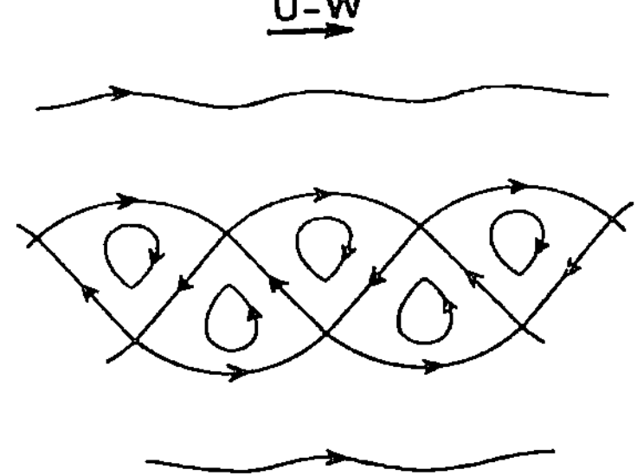

Fig:2   A vortex street seen by an observer accompanying the eddies   [26]

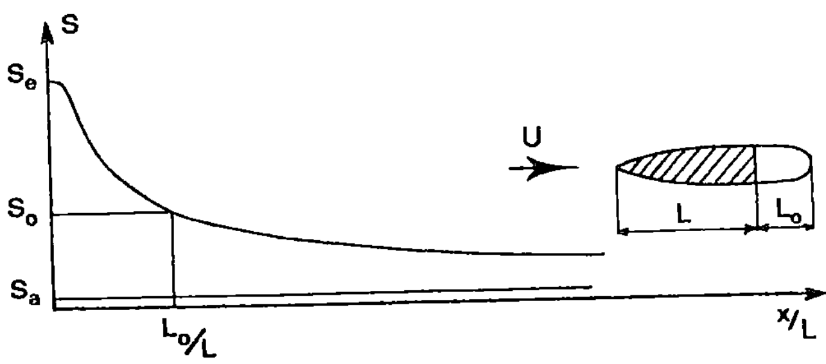

Fig:3   Formation of the vortex street   [12,26]

M = .70

α = 4°

R = 10^6

$M = .70$

$\alpha = 4^{\circ}$

$R = 10^6$

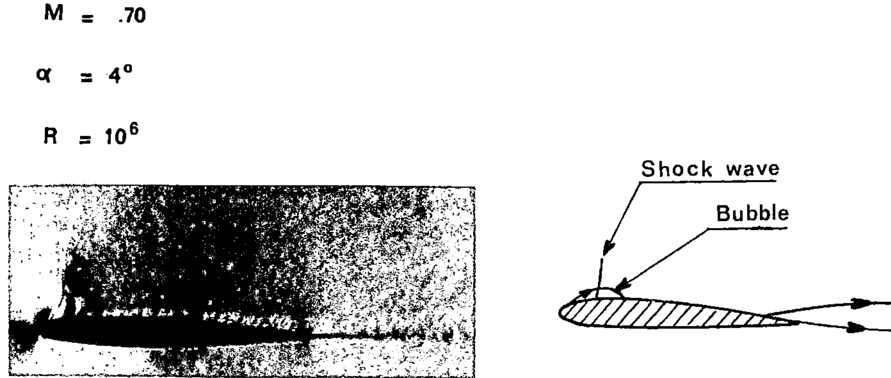

Fig :4 . Supercritical flow around an airfoil [26]

$M = .30$

$\dfrac{U H}{\nu} = .8 \; 10^5$

$\Delta t \simeq 200 \, \mu s$

One period of the

vortex street

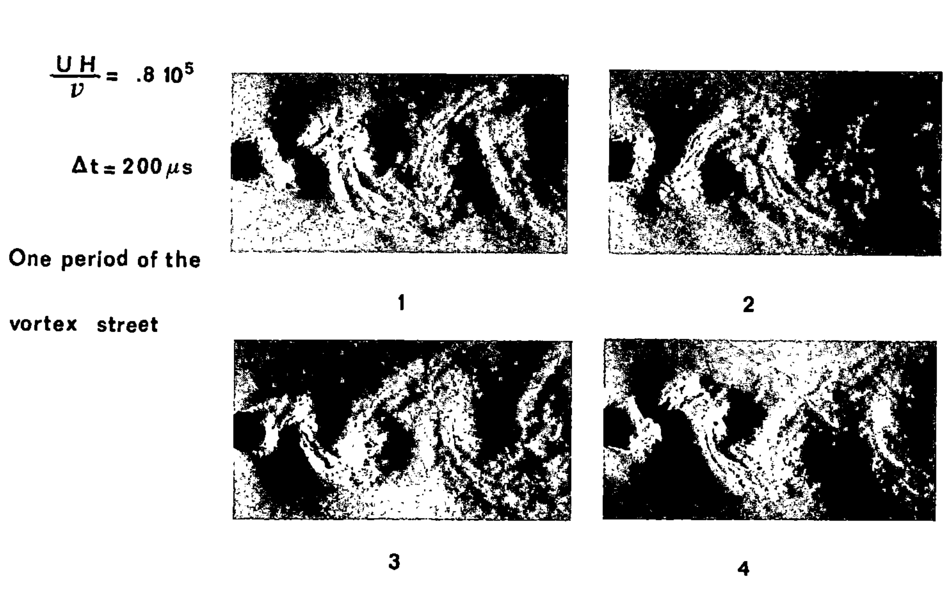

Fig:5 Cylinder . High speed visualization [26]

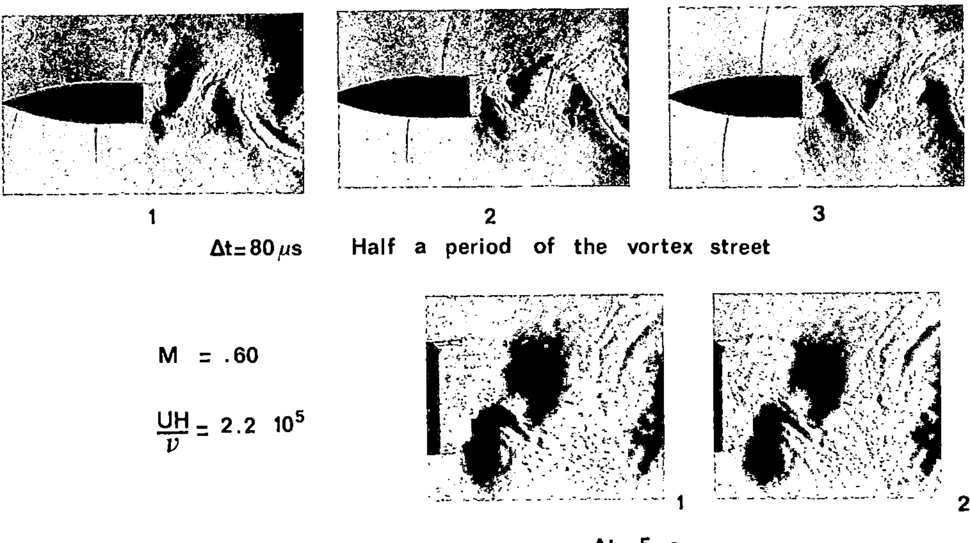

Fig:6   Base flow. High speed visualization   [25]

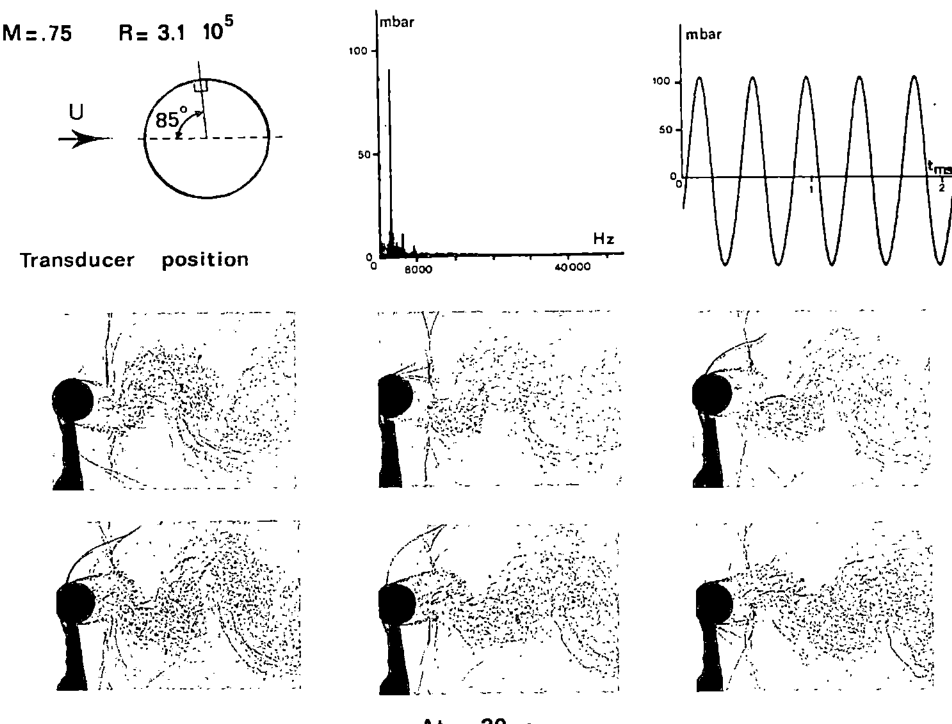

Fig:7   Cylinder. High speed visualization and pressure measurements   [28]

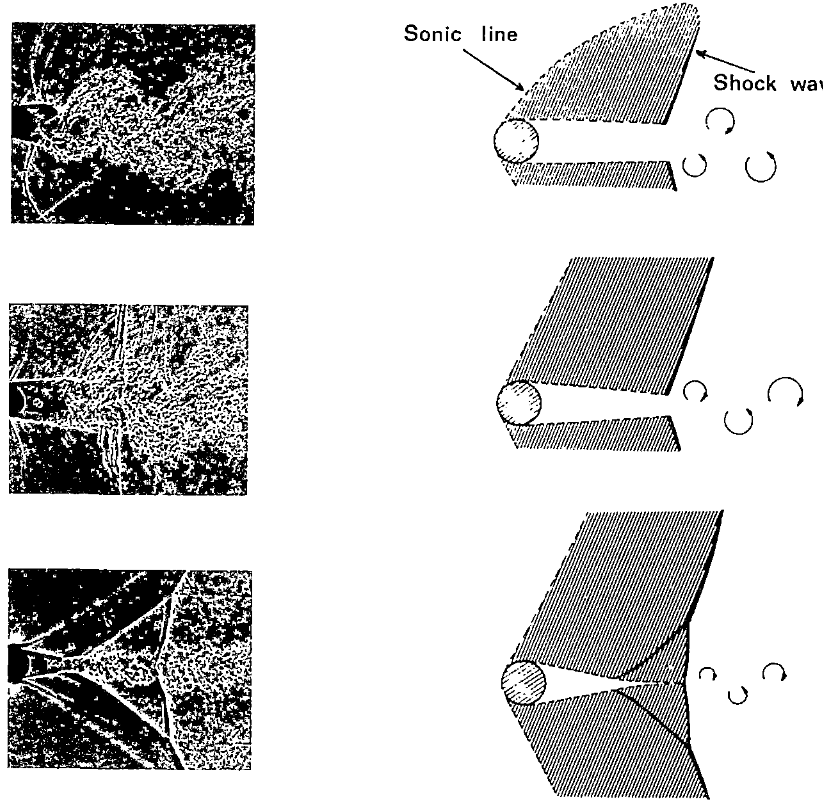

Fig:8 Cylinder. Different supercritical flow regimes [27]

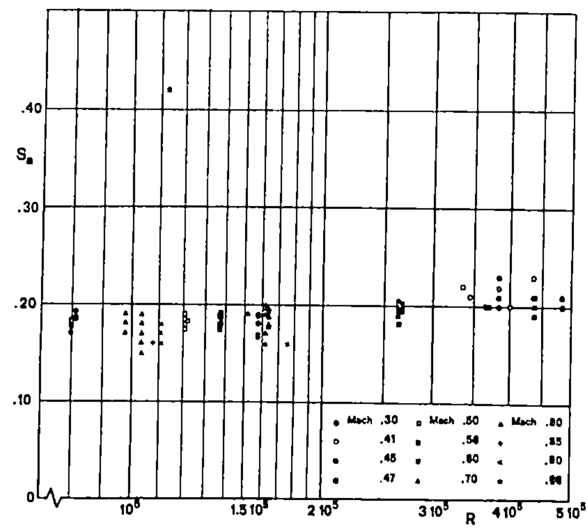

Fig :9 Cylinder. Vortex street Strouhal number [25]

VORTEX SHEETS AND CONCENTRATED VORTICITY
A VARIATION ON THE THEME OF ASYMPTOTIC MODELLING IN FLUID MECHANICS

by

J.P. Guiraud[+], R. Zeytounian[x]

Office National d'Etudes et de Recherches Aérospatiales
29 Avenue de la Division Leclerc, 92320 CHATILLON (FRANCE)

Introduction

Some twenty five years ago, with the works of Kaplun [1954] [1957], Kaplun and Lagerstrom [1957], Proudman, Pearson [1957], asymptotic techniques gave a new impetus on research in theoretical fluid dynamics. Ten years later, a much more powerful revival was possible thanks to the dramatic influence of high speed computers and its companion the numerical analysis of fluid flow problems. During the early times, asymptotic techniques were mostly used in order to get approximate solutions in closed form. Perhaps, of more significance for the progress of understanding and also of research, was the use of asymptotic techniques in order to settle, on a rational basis, a number of approximate models which were, often much earlier, derived by "ad-hoc" procedures. One of the most famous examples is Kaplun's [1954] celebrated paper on boundary layer theory, which gave to some fifty years of boundary layer research a firm theoretical basis. Already, from this early example, it might have been clear that asymptotic techniques were well suited to derive mathematical models amenable to numerical treatment rather than to obtain closed form solutions. However, at the time, numerical fluid dynamics was almost non-existent due to the lack of high speed computers.

It is now evident that asymptotic techniques provide very powerful tools in the process of constructing mathematical models for problems which are stiff, from the point of view of numerical analysis. A whole class of such stiff problems is encountered in the so-called viscous-inviscid strong interactions. The asymptotic modelling of this class of problems, the so-called triple-deck theory, was created simultaneously in great-Britain (Stewartson, Williams [1969]) and in Soviet-Union (Neiland [1969]), while numerical fluid dynamics was in an explosive expansion. The triple deck model is now a basic one in the field of numerical fluid dynamics, as applied to viscous-inviscid interactions (see Adamson , Messiter [1980] for a recent review of part of the subject) even when extended to turbulent boundary layers (see again Adamson , Messiter [1980]) a field of research where more direct and more traditional approaches (Le Balleur [1981]) might appear as better suited.

It is our opinion that asymptotic modelling will remain for many years, or even decades, a quite powerful tool in deriving mathematical models for numerical fluid dynamics research. By mathematical, we mean that the model under consideration should be formulated as a "reasonable well posed" initial and/or boundary value problem for ordinary or partial differential equations (or more general type of equations). We do not mean that there exists a rigorous mathematical theory of the problem at hand. By amenable to numerical analysis we mean that the problem at hand may be investigated by a numerical code

+ Laboratoire de mécanique théorique, Université P. et M. Curie, Tour, Place Jussieu 75230 Paris Cedex 05, France.

x UER de mathématiques pures et appliqués, Université des Sciences et Techniques, de Lille 1, BP 36, 59560 Villeneuve d'Asq, France.

constructed through application of some rational technique of finite difference, finite element, Galerkin,... type, even if there is no proof of convergence of the finite dimensional approximation to the continuous solution. Incidently, the reader interested in the general concept of mathematical modelling as applied to fluid dynamics is referred to Landahl [1981].

In the present paper, we intend to illustrate our interpretation, as defined previously, of asymptotic modelling, with three examples all borrowed from vortex flow configurations to which the present volume is devoted. Computations of flow involving concentrated vorticity is a very active field of research in numerical fluid dynamics as applied to aerodynamics (see Rehbach [1978] for examples and Leonard [1980] for a review of numerical codes relying strongly on vorticity concepts).

Our first example is concerned with the modelling of flows involving slender vortex filaments. The old problem of motion of vortex rings was revisited some ten years ago in reason of the strong necessity of mastering the dynamics of trailing vortices issued from big aircrafts. It was then discovered by Ting and Tung [1965] and Widnall, Bliss and Zalay [1971] that this problem was quite well suited for asymptotic techniques. As a matter of fact it would be hopeless to attack the dynamics of trailing vortices through direct numerical simulations. Here we present briefly this basic work, from the point of view of computation of flow fields involving slender vortex filaments, and we stress that the work should be extended to compressible flow.

Our second topic concerns flow structures which involve very many closely spaced vortex sheets. Here we rely on some work of our own (Guiraud φ Zeytounian [1977, 1979 a, 1979 b]) that we extend here for the first time. We put the emphasis on the point of view that the theory offers a mathematical model which allows to avoid the very stiff problem of having to follow the sheets without mixing them, a problem which is faced by any numerical code devised for computing cores of rolled vortex sheets.

Our last topic concerns the construction of model equations for cores of either slender or highly swirling vortex flows. This research is motivated by problems of flow in vortex chambers (for which we refer to Lewellen [1962]) or in geophysical vortices (for which we refer to Morton [1966]). Here we investigate model equations derived through various limiting processes.

2. Vortex filaments

The first point that we want to discuss is the motion of vortex filaments. These are regions of concentrated vorticity, embedded in an otherwise irrotational flow, which are "tube-like" and very slender. The numerical treatment of such flows enters into the category of stiff problems because, in order to provide a reliable computation, one would have to use a very tiny mesh grid in the vicinity of the filament. On the other hand, the grid should neither be of the Lagrange nor of the Euler type. As a matter of fact, it should be Lagrange-like concerning the gross motion of the whole vortex filament, but it should be Euler-like concerning the motion inside the filament. Obviously, one has to deal with a kind of problem amenable to asymptotic modelling. There is at least one small parameter built into the problem and this is the slenderness ratio ϵ , which is the ratio of the mean diameter of the cross section of the filament to the radius of curvature of a fictitious line, to which the region of concentrated vorticity is expected to collapse under the limiting process $\epsilon \rightarrow 0$, but that, during the process, the circulation around the filament remains finite. Asymptotic modelling, as applied to the problem at hand, amounts to use asymptotic techniques in order to devise a mathematical model of the flow as a whole, amenable to a numerical treatment, from which stiffness has disappeared. Matched asymptotic techniques (MAT) seem to be well suited for this goal. As a matter of fact, they should provide the numerical fluid dynamicist with two separate models, each one being amenable to numerical tratment without stiffness. The gross flow model concerns the flow away from, but up to a vicinity of the vortex filament, while the core flow model has to deal with the flow within the so-called core of the vortex.

The gross flow model, to leading order, is one of flow without concentrated vorticity. Any numerical code, suitable for dealing with inviscid flow, would be appropriate, provided it had, built in, a subroutine appropriate to the singular type of behaviour of the flow variables when one gets close to the limiting line vortex $\mathcal{L}$. By this we mean the line (s) to which the vortex filament (s) collapse under the limiting process $\epsilon \to o$, circulation $\to \Gamma$. The analytic apparatus underlying the subroutine of motion close to the line vortex should be a by-product of the process of matching the two expansions which are at the heart of the two models. Here we stress that the gross flow model is especially simple when the flow is incompressible and irrotational outside from the region of concentrated vorticity. The model has two main ingredients. The first one is a code for computing the velocity potential ϕ , subjected to the Laplace's equation $\Delta\phi = 0$ and to any kind of boundary conditions which may be suited to the velocity field

$$(1) \qquad \vec{u}\,(t,\vec{x}) = \vec{\nabla}\phi - \sum_{i=1}^{N} \frac{\Gamma_i}{4\pi} \int_{\mathcal{L}_i} \frac{(\vec{x}-\vec{y}_i)\wedge d\vec{y}_i}{|\vec{x}-\vec{y}_i|^3} ,$$

where the sum $\sum$ is over all the line vortices embedded within the flow. The second ingredient is an algorithm for the time evolution of each of the $\mathcal{L}_i$. The search for such an algorithm is an old and very long story. It was given a strong impetus at the end of the XIXth century by the work of W. Thomson [1867] (alias Lord Kelvin) who suggested that the flow configuration known as the vortex ring, might be the basic constituent of matter. The idea has long been given up, but the work on vortex rings, initiated by J.J. Thomson [1883], following the fundamental paper of Helmholtz [1858] has been revived many times, especially during the last decade. It was already known to Kelvin [1867], but proved by Hicks [1885], that a torous of radius R and circular cross section of radius a, filled with uniform axial vorticity, is a steady flow configuration which moves, within an irrotational flow at rest at infinity, with a constant velocity of translation, perpendicular to the plane of the ring, given, asymptotically, by

$$(2) \qquad V = \frac{\Gamma}{4\pi R} \left\{ \mathrm{Log}\ \frac{8R}{a} - \frac{1}{4} + O\left(\frac{\Gamma a}{R^2}\ \mathrm{Log}\ \frac{a}{R}\right)\right\}.$$

The mathematical proof of the existence of a steady configuration, in the form of a vortex ring in inviscid incompressible flow, which is asymptotically toroidal, with uniform vorticity and velocity given by (2), in the case when $a/R \ll 1$, is a quite involved one and was given by Fraenkel [1970] [1972]. The formula (2) may be derived by applying to the vortex ring as a whole, some integral invariants of the vorticity distributions. That was the way followed by Hicks [1885] as reported by Lamb [1932] (paragraph 161-164) and fully exploited by Saffman [1970].

Although this way of approaching the problem, of finding an algorithm
for the time evolution of the line vortex, is a very illuminating one, as
shown by the work of Moore and Saffman [1972], on vortex filaments of general
shape, we think that its profound significance is most efficiently stressed
when it is viewed as a by-product of the matching technique. This point of
view was probably argued for the first time by Ting and Tung [1965] and reasses-
sed by Widnall, Bliss and Zalay [1971] who corrected a minor error in Tung &
Ting [1967].   For details we refer to Callegari & Ting [1979] from which
we state the result concerning the velocity $\vec{u}_{\ell_i}(t, \vec{x})$ of the point $\vec{x}$,
located at time t on the ith line vortex ℓ_i, namely

$$(3) \quad \begin{cases} \vec{u}_{\ell_i} \cdot \vec{\tau}_i = 0 \quad , \quad \vec{u}_{\ell_i} \cdot \vec{n}_i = \vec{V}_i \cdot \vec{m}_i \, , \\[2mm] \vec{u}_{\ell_i} \cdot \vec{b}_i = \vec{V}_i \cdot \vec{b}_i + \dfrac{\Gamma_i}{4\pi R_i} \, \mathrm{Log}\, \dfrac{R_i}{\alpha_i} + \vec{F}_i(t,\vec{x}) \, . \end{cases}$$

According to these formulae, a_i is the radius of the cross-section of the vortex filament. Such a section is asymptotically circular and centred on the
line vortex ℓ_i, to which is associated a Serret–Frenet triad $\vec{\tau}_i, \vec{m}_i, \vec{b}_i$,
in such a standard way that the radius of curvature R_i of ℓ_i be positive. The
vector $\vec{V}_i$ is found through a limiting process

$$\vec{V}_i = \vec{\nabla}\phi - \sum_{j \neq i} \frac{\Gamma_j}{4\pi} \int_{\ell_j} \frac{(\vec{x} - \vec{y}_j) \wedge d\vec{y}_j}{|\vec{x} - \vec{y}_j|^3} -$$

$$(4) \quad - \lim_{n/R_i \to 0} \frac{\Gamma_i}{4\pi} \left\{ \int_{\ell_i} \frac{(\vec{z} - \vec{y}_i) \wedge d\vec{y}_i}{|\vec{z} - \vec{y}_i|^3} + \frac{2}{n}\vec{f} + \right.$$

$$\left. + \frac{1}{R_i}\left(\mathrm{Log}\,\frac{R_i}{n}\right)\vec{b}_i + \frac{\cos\varphi}{R_i}\vec{f} \right\} \, ,$$

where

$$(5) \quad \begin{cases} \vec{z} = \vec{x} + n\,\vec{e}(t,\vec{x},\varphi) \quad , \quad \vec{f} = \vec{\tau}_i \wedge \vec{e} \, , \\[2mm] \vec{e} = \vec{m}_i \cos\varphi + \vec{b}_i \sin\varphi \, , \end{cases}$$

while the function $\vec{F}_i(t,\vec{x})$ is the only part of the algorithm through which
the internal structure of the vortex is felt. It should be stressed that in
the whole algorithm for evolution of the line vortices, the dynamics enters
only through $\vec{F}_i(t,\vec{x})$. All other ingredients are purely kinematic ones.

Curiously enough, to the best of our knowledge, slender vortex filaments
embedded in compressible flows have attracted little attention. When the colloquium came to the author's knowledge it was planned to undertake a research
program on this question. When it was known that the manuscript were planned
to be printed in advance it was clear that this research had to be delayed
in order to prepare a manuscript with the material at our disposal. We would
be happy if this lecture might convince people engaged in research of the importance of asymptotic modelling and if somebody were to carry to completion
our initial program concerning compressible vortex filaments.

We have focussed our attention to asymptotic modelling oriented towards
numerical computing because this facet of asymptotic modelling deserves to be
emphasized. However, there is a lot of works that rely on asymptotic techniques in order to get results in almost closed form or with a light support
of computer codes. Of such a nature is the work on stability of vortex rings
or of vortex filaments, either isolated or paired as in trailing vortices of
big aircraft , but we shall not discuss this point and refer for that to
Saffman&Baker [1979].

We close this section by referring to Moore [1978] who investigated a somewhat analogous situation for plane motion. A vortex filament is then a thin sheet carrying vorticity, while the flow outside is irrotational. Let a be the thickness of the sheet and R a typical radius of curvature of the mean line ; let ω be the vorticity within the sheet. The limiting process under consideration is $\varepsilon = a/R \to 0$, $\omega \to \infty$ in such way that $\varepsilon\omega \to K$. There is an analogue of the Biot and Savart's formula (1); it is known as the Birkhoff's formula and it may be applied to obtain the material velocity of the line vortex when the shape of this line and the distribution of vorticity along it is known. Let Γ be the circulation attached to a part of the line vortex beginning at one extremity of it and ending in a material point, and let z be the complex position $z = x + i y$ corresponding to this point. The Birkhoff formula may be written as

$$(6) \qquad \frac{\partial z^*}{\partial \Gamma} = - \frac{i}{2\pi} \int_0^\Gamma \frac{d\Gamma'}{z(t,\Gamma) - z(t,\Gamma')} \; , \qquad z^* = x - i y \; ,$$

where the principal value is to be understood. What Moore [1978] proves is that this formula is obtained in a process of matching an asymptotic representation of the velocity field at O (1) (with R as unit) distances from the line vortex, to another one valid at distances $\Omega(\varepsilon)$. The analysis provides also an $O(\varepsilon)$ correction to (6).

3. Vortex sheets and related flow structures

Vortex sheets are one of the basic ingredients of inviscid flows. Contrary to vortex filaments which cannot be viewed as vorticity concentrated on a line, but must be considered as having a core of small but finite diameter, vortex sheets may be considered as two-dimensional surfaces of zero thickness carrying a truly concentrated vorticity. This may be enlightened by writing the vorticity as follows :

$$(1) \qquad \vec{\omega} = \vec{\omega}^s + \delta_\Sigma \vec{n} \wedge [\![\vec{u}]\!] = \vec{\omega}^s + \vec{\omega}^c \; , \qquad \vec{\omega}^c . \vec{n} = 0 \; ,$$

where $[\![\vec{u}]\!]$ is the jump in velocity across the vortex sheet Σ with unit normal $\vec{n}$, and δ_Σ is the Dirac distribution uniformly spread over Σ . The jump in $\vec{u}$ is defined as

$$(2) \qquad [\![\vec{u}]\!] = \vec{u}^+ - \vec{u}^- \qquad , \qquad [\![\vec{u}]\!] . \vec{n} = 0 \; ,$$

where + stands for the side of Σ which is encountered in second when Σ is crossed in the sense of $\vec{m}$. Let us set $f_m = \frac{1}{2}(f^+ + f^-)$ for the mean value of f on the two sides of Σ , the dynamics leads to the following equations

$$(3) \qquad \begin{cases} \dfrac{D_m [\![\vec{u}]\!]}{D t} + \left([\![\vec{u}]\!] . \vec{\nabla} \vec{u}_m \right)_T + \left[\dfrac{1}{\rho} \right] \vec{\nabla}_T p = 0 \; , \\[2ex] \dfrac{D_m S}{D t} + [\![\vec{u}]\!] . \vec{\nabla} S_m = 0 \; , \end{cases}$$

where p is the pressure, ρ the density and S the entropy, while $D_m/Dt = \partial/\partial t + \vec{u}_m . \vec{\nabla}$ is a derivative along the sheet corresponding to the mean velocity $\vec{u}_m$.

We clearly see that, as long as compressible fluid is concerned, concentrated vorticity is not the whole matter ; we should add a concept of concentrated baroclinicity vector $\vec{B} = \vec{\nabla} S$ by considering that, along any surface orthogonal to $\vec{B}$, variation of p and ρ , at constant time, are related by a barotropic relation. Then, by writing, in a way analogous to (1)

$$(4) \qquad \vec{B} = \vec{B}^s + \oint_{\Sigma} [\![S]\!]\, \vec{n} = \vec{B}^s + \vec{B}^c ,$$

we exhibit a concentrated baroclinicity $\vec{B}_c$ which is orthogonal to the concentrated vorticity. We may state that

$$(5) \qquad [\![\tfrac{1}{\rho}]\!] = \frac{\partial \widetilde{T}}{\partial \rho}\, [\![S]\!] ,$$

where $\frac{\partial \widetilde{T}}{\partial \rho}$ is calculated at ρ and $\widetilde{S} = \alpha S^+ + (1-\alpha) S^-$ with $0 < \alpha < 1$.

The situation that we want to describe is one in which there are very many vortex sheets closely spaced between each other. We may mention two flow configurations in which this occurs. The first one is the core of a highly rolled vortex sheet, while the second concerns the set of trailing vortex sheets which are formed at the trailing edges of the blades in a row of an axial turbomachine (Guiraud, Zeytounian [1978]). We assume that there is a small parameter built in the flow, which is the ratio of the spacing between two consecutive sheets to the width of the region covered by the sheets. We set C for this parameter and, according to Guiraud and Zeytounian [1979 b], we call it a closeness parameter. The purpose of asymptotic modelling in this situation is to derive a model which avoids the stiff problem of computing the flow with a numerical code capable of capturing many sheets. The problem is one of multiple scaling in the terminology of asymptotics. A version of the multiple scale technique, especially suited for this kind of problem was devised by Guiraud and Zeytounian [1977, 1979 a, 1979 b] with the purpose of describing rolled vortex sheets, but its scope is more general as it may be seen from the analysis to follow. We start from the assumption that the velocity $\vec{u}$, pressure ρ, density ρ, entropy S, all suitably made non-dimensional, are functions of time t, position $\vec{x}$ and of one fast variable $C^{-1}\chi(t,\vec{x})$, with $\vec{\nabla}\chi$ approximately orthogonal to the sheets. We use the notation

$$(6) \qquad \mathcal{U} = (\vec{u},\, \rho,\, \rho,\, S)^{\mathsf{T}},$$

and we set

$$(7) \qquad \mathcal{U}(t,\vec{x}) = \mathcal{U}^{*}\!\left(t,\vec{x},\, C^{-1}\chi(t,\vec{x})\right).$$

There are two ingredients in the technique used. The first one is a formal expansion

$$(8) \qquad \mathcal{U}^{*} = \mathcal{U}^{*}_{o} + C\,\mathcal{U}^{*}_{1} + \cdots ,$$

while the second one is the obvious observation that, setting

$$(9) \qquad \begin{cases} \theta = \dfrac{\partial \chi}{\partial t} = \theta_o + C\,\theta_1 + \cdots , \\[2mm] \vec{k} = \vec{\nabla}\chi = \vec{k}_o + C\vec{k}_1 + \cdots , \end{cases}$$

we have

$$(10) \qquad \begin{cases} \dfrac{\partial \mathcal{U}}{\partial t} = C^{-1}\theta\,\dfrac{\partial \mathcal{U}^{*}}{\partial \chi} + \dfrac{\partial \mathcal{U}^{*}}{\partial t} , \\[3mm] \vec{\nabla}\mathcal{U} = C^{-1}\dfrac{\partial \mathcal{U}^{*}}{\partial \chi}\,\vec{k} + \vec{\nabla}\mathcal{U}^{*} . \end{cases}$$

By substituting (8) (9) (10) into the equations of motion, we get, at zeroth order, a set of equations from which we conclude that, provided $\chi =$ const is not a Mach wave, we must have that $\theta_o + \vec{k}_o \cdot \vec{u}_o^{*}$ does not depend on χ. On the other hand, by specifying that χ is constant on the vortex sheets we get

36

$$(11) \qquad \theta_o + \vec{k}_o \cdot \vec{u}_o^* = 0 \ .$$

Then, the only conclusion that may be derived from the very degenerate equations at zeroth order is the following

$$(12) \qquad \vec{k}_o \cdot \frac{\partial \vec{u}_o^*}{\partial \chi} = \frac{\partial \pi_o^*}{\partial \chi} \ ,$$

the first one of which is in agreement with (11).

Let us consider the vorticity $\vec{\omega}$. From the very definition we get

$$(13) \qquad \vec{\omega}^* = C^{-1} \vec{k} \wedge \frac{\partial \vec{u}^*}{\partial \chi} + \vec{\nabla} \wedge \vec{u}^* \ ,$$

so that, if we enforce that $\vec{\omega}^*$ is 0 (1), then, as was the case in the study of rolled-up vortex sheets, we have

$$(14) \qquad \vec{k}_o \wedge \frac{\partial \vec{u}_o^*}{\partial \chi} = 0 \ ,$$

and this, in turn, enforces that

$$(15) \qquad \frac{\partial \rho_o^*}{\partial \chi} = \frac{\partial S_o^*}{\partial \chi} = 0 .$$

The model of rolled vortex sheets is recovered by assuming that u^* is 3π-periodic with respect to χ ; then, by writing the equations at order one, it may be derived that $\vec{V} = \vec{k}_o \wedge \vec{u}^*$ and $\Sigma = S_1^*$ are solutions of

$$(16) \quad \left\{ \begin{aligned} & \frac{\partial \vec{V}}{\partial t} + \vec{u}_o^* \cdot \vec{\nabla} \vec{V} - \vec{V} \cdot \vec{\nabla} \vec{u}_o^* + (\vec{\nabla} \cdot \vec{u}_o^*) \vec{V} + \\ & \qquad + \frac{\vec{k}_o \cdot (\vec{\nabla} \wedge \vec{u}_o^*)}{|\vec{k}_o|^2} \ \vec{k}_o \wedge \vec{V} + \frac{1}{\gamma M^2} \ \frac{\vec{k}_o \wedge \vec{\nabla} \pi_o^*}{\gamma \rho_o^*} \ \Sigma = 0 \ , \\ & \frac{\partial \Sigma}{\partial t} + \vec{u}_o^* \cdot \vec{\nabla} \Sigma + \frac{\vec{k}_o \wedge \vec{\nabla} S_o^*}{|\vec{k}_o|^2} \cdot \vec{V} = 0 \ . \end{aligned} \right.$$

The first of these equations is a consequence of the equation of vorticity. The important point is that χ does not occur in (16). They are a set of ordinary differential equations along the trajectories of the velocity field at zeroth order approximation $\vec{u}_o^*$. One may set

$$(17) \qquad (\vec{V}, \Sigma) = \mathcal{F}_o(t^o, t) \, (\vec{V}^o, \Sigma^o) \ ,$$

stating that the field $\vec{V}(t, \vec{x})$, $\Sigma(t, \vec{x})$ at time t is known once the field $\vec{V}^o(\vec{x})$, $\Sigma^o(\vec{x})$ at time t^o is given. In principle, the linear operator $\mathcal{F}_o(t^o, t)$ can be computed from a numerical code when the zeroth order field $\vec{u}_o^*, \pi_o^*, S_o^*$ is known. We may set at once

$$(18) \qquad (\vec{k}_o \wedge \vec{u}_1^*, S_1^*) = \mathcal{F}_o(t^o, t) \left(\vec{k}_o \wedge \vec{u}_1^{*o}, S_1^{*o} \right) \ ,$$

and the whole of the dependency on χ appears through $\vec{u}_1^{*o}$ and S_1^{*o} , the values of $\vec{u}_1^*$ and S_1^* at t^o.

A number of configurations are allowed with this description. We present two of them. The first one which corresponds to the left part of the figure, consists in vortex sheets of the same intensity situated at $\chi = (2h+1)\pi$. If the flow is incompressible and irrotational between the sheets, $\vec{k}_o \wedge \vec{u}_1^*$ is linear in χ between the sheets and 2π- periodic. Of course, there is no entropy. If we set $[\![\vec{u}_1^*]\!]$ for the jump of $\vec{u}_1^*$ across any of the sheets one has

$$\vec{k_0}\cdot[\![\vec{u_1}^{*}]\!] = 0 \quad \text{and}$$

$$(19) \qquad \vec{u_1}^{*} = -\frac{\chi}{\pi}[\![\vec{u_1}^{*}]\!] \quad , \quad -\pi < \chi < \pi \; .$$

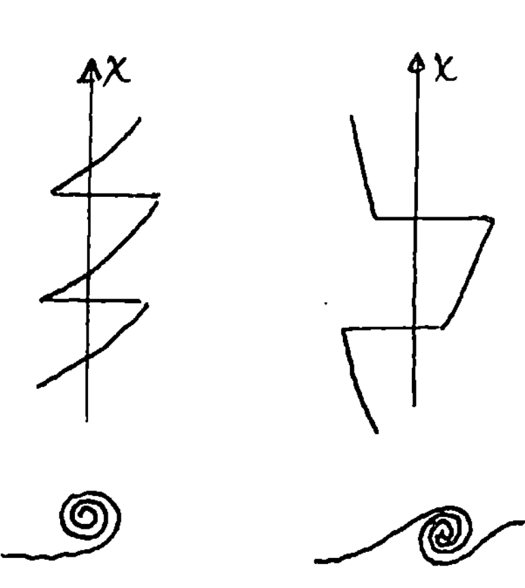

This definition of $\vec{u_1}^{*}$ may be extended to all values of χ by a periodicity argument. It may be checked that the first of (16) may be reduced to the first of (3). Consider now the second configuration, which corresponds to the right part of the figure. We have two counter rotating spirals and two discontinuities in one period. If the flow is again incompressible and irrotational between the sheets one may write, corresponding to (19), the following formula

$$(20) \qquad \vec{u_1}^{*} = \begin{cases} |\vec{k_0}|^{-2}\,\chi\;\vec{k_0}\wedge(\vec{\nabla}\wedge\vec{u_0}^{*}) + \vec{A} & , \quad -\pi < \chi < \chi_0 \;, \\[2mm] |\vec{k_0}|^{-2}\,\chi\;\vec{k_0}\wedge(\vec{\nabla}\wedge\vec{u_0}^{*}) + \vec{B} & , \quad \chi_0 < \chi < \pi \;, \end{cases}$$

with

$$(21) \qquad \pi(\vec{A}+\vec{B}) + \chi_0(\vec{A}-\vec{B}) = 0 \;,$$

if one insists that the mean value of $\vec{u_1}^{*}$ over $(-\pi,\pi)$ be zero as was done in (19). Two unknowns occur in (20). One of them is χ_0 and it is related to the ratio of two consecutive spacings ; the second is $\vec{A}-\vec{B}$ and it is related to the ratio of intensities of vorticity concentrated on two consecutive sheets. The fact that there must be unknowns in the function $\vec{u_1}^{*}(\chi)$ is made quite clear if one goes back to (18). Let us associate, at time t^{0}, to each point, in the region of concentrated vorticity, a basis $\{\vec{e_1^{0}}, \vec{e_2^{0}}\}$ for vectors in the plane normal to $\vec{\nabla}\chi$. Let us set, at time t

$$(22) \qquad (\vec{e_i}, G) = \mathcal{F}_0(t^{0}, t)\,(\vec{e_i^{0}}, 1),$$

then one may state, from (18), that

$$(23) \quad \begin{cases} \vec{k_0}\wedge\vec{u_1}^{*} = \varphi_1(\chi)\,\vec{e_1}(t,\vec{x}) + \varphi_2(\chi)\,\vec{e_2}(t,\vec{x}) \;, \\[2mm] S_1^{*} = \varphi_3(\chi)\,G(t,\vec{x}) \;, \end{cases}$$

with three arbitrary functions of χ . If one considers that, once entered into the region of concentrated vorticity, a fluid particle penetrates deeper and deeper into the rolled core as the time goes on, one realizes that the functions $\varphi_1, \varphi_2, \varphi_3$ represent nothing else than a memory of vorticity and baroclinicity in the early times of the history, when a given blob of fluid entered into the rolled region .

It should be clear, from now, that once the zeroth order field $\vec{u_0}^{*}, \vec{\eta_0}^{*}, \rho_0^{*}$ has been computed, from a numerical code for example, formulae (22), (23)

allow to continue the fine scale description deeper and deeper into the rolled
region, provided it is known in some early stage of the rolling process.
Asymptotic modelling, for the problem at hand, has accomplished what it was
devised for, namely to provide a substitute to the numerical code or, more
precisely, to a part of it, when this part becomes too stiff. An application
of this may be found in Huberson [1980]. Here we review rapidly some applications for which the basic field of flow is known analytically. Using Hall's
[1961] solution for the rotational conical axisymmetric incompressible flow,
in the vicinity of the axis, Guiraud and Zeytounian [1977] rederived Mangler $\mathcal{A}$
Weber's [1967] solutions for the core of a rolled leading edge vortex.
Using Brown's [1965] compressible version of Hall's [1961] solution they
rederived Brown $\mathcal{A}$ Mangler [1967] extension to compressible barotropic flow
of the Mangler $\mathcal{A}$ Weber [1967] one. The versatility of the technique is demonstrated by two other closed form solutions. The first one, by Guiraud $\mathcal{A}$
Zeytounian [1977] concerns the so-called Kaden's problem. At time $t = 0$ one
is given an irrotational incompressible stealy two-dimensional flow round a
straight edge, this last one being removed instantaneously and replaced by
a vortex sheet which rolls-up. If the edge is along $x < 0$, $y = 0$ and if the
complex initial velocity is $u^0 - iv^0 = i\frac{\gamma}{2}(x + iy)^{-1/2}$, in
dimensionless form, the sheet is found to be

$$(24) \quad \begin{cases} x = (\gamma t)^{2/3}(a + \xi \cos\theta), \\ y = (\gamma t)^{2/3}(b + \xi \sin\theta), \\ \theta = \theta_0 - \beta \xi^{-2/3}\{1 + \mathcal{F}(\theta, \xi)\}, \end{cases}$$

where a, b, β, θ_0 are numerical constants while $\mathcal{F}$ is expanded, near $\xi = 0$
with respect to increasing [non integer] powers of ξ , namely

$$(25) \quad \mathcal{F} \simeq \sum_{k=1}^{K} \xi^{n_k} \mathcal{F}_k(\theta) \; ; \quad m_1 = 0.579 \; ; \; n_2 = 1.054 \; ; \; m_3 = 1.153$$

The function $\chi(t, x, y)$ is found to be

$$(26) \quad \chi = \theta - \theta_0 + \beta \xi^{-2/3}\{1 + \mathcal{F}(\theta, \xi)\}.$$

The velocity field $\vec{u}_0^*$ is stated in terms of its radial u_0^x and tangential v_0^*
components as

$$(27) \quad \begin{cases} u_0^\xi = \gamma\beta(\gamma t \xi)^{-1/3} \mathcal{U}(\theta, \xi), \\ v_0^* = -\gamma\beta(\gamma t \xi)^{-1/3}\{1 + \mathcal{V}(\theta, \xi)\}, \end{cases} \quad \mathcal{U} = \sum_k \xi^{m_k} \mathcal{U}_k(\theta), \quad \mathcal{V} = \sum_k \xi^{n_k} \mathcal{V}_k(\theta).$$

The functions $\mathcal{F}_k(\theta)$, $\mathcal{U}_k(\theta)$, $\mathcal{V}_k(\theta)$ are simple trigonometric functions.
The solution for the sheet was known to Kaden [1931] but only as far as the
leading term. The form given above for χ is implicit in the work of Moore
[1975] who gave the expansion up to $O(\xi^{m_1})$ while Guiraud $\mathcal{A}$ Zeytounian [1977]
proceeded to $O(\xi^{m_3})$. One more closed form solution was given by Guiraud
[1977] who proved that any vortex filament, according to §2 above, may be the
support of a vortex sheet tightly wound around the line vortex which is the
support of the filament.

All the closed form solutions referred to previously arise from the fact
that the closeness parameter C is related often, but not always, by $C = \varepsilon^\lambda$
to a slenderness parameter ε with respect to which $\vec{u}_0^*$ may be expanded. However,
we emphasize here that getting closed form solutions should not be considered
as the ultimate goal of the analysis. In our opinion, the analysis should
rather be considered as a model allowing to generate a flow with closely
spaced vortex sheets from another one which does not involve such a feature.

We come now to a more general situation by relaxing the condition that, in (13), $\vec{\omega}^*$ remain 0 (1) when $c \to 0$. On the other hand, we accept that the expansion of $\vec{\omega}^*$ starts with an $O(c^{-1})$ term and we set

$$(28) \qquad \vec{\omega}^* = c^{-1}\, \vec{\Omega}^* \, ,$$

from which we get

$$(29) \qquad \vec{\Omega}^* = \vec{k} \wedge \frac{\partial \vec{u}^*}{\partial \chi} + c\, \vec{\nabla} \wedge \vec{u}^* \, .$$

To zeroth order, the equations of motion are such highly degenerated ones that the sole conclusion that we are allowed to deduce from them is that (12) holds true. At the same time, we remove the condition that $\mathcal{U}^*$ be periodic with respect to χ . The only condition that we preserve is that the terms in (8) are well ordered, which amounts to enforcing that $\mathcal{U}_1^*$ remains 0 (1) even when χ grows without bound. This is a basic ingredient in multiple scaling techniques. Let $f^*(\chi)$ be any function of $t, \vec{x}, \chi$ where we have omitted the dependency on t and $\vec{x}$; we set

$$(30) \qquad \begin{cases} \overline{f^*} = \lim_{\substack{\chi_1 \to -\infty \\ \chi_2 \to +\infty}} \frac{1}{\chi_2 - \chi_1} \int_{\chi_1}^{\chi_2} f^*(\chi)\, d\chi \, , \\[2mm] f^* - \overline{f^*} = \widetilde{f^*} \, . \end{cases}$$

We say that $\overline{f^*}$ is the mean of f^* while $\widetilde{f^*}$ is its fluctuation. Assume that the equations of motion have been written in dimensionless conservative form, namely

$$(31) \qquad \frac{\partial \mathcal{V}}{\partial t} + \frac{\partial \mathcal{F}_K(\mathcal{V})}{\partial x_K} = 0 \, ,$$

where summation over $K = 1, 2, 3$ is implied. For definiteness we state that

$$(32) \qquad \mathcal{V} = (\, \rho, \, \rho u_1, \, \rho u_2, \, \rho u_3, \, \rho S\,)^T \, ,$$

and leave to the reader the writing out of the $\mathcal{F}_K$. From (6) to (10) we get

$$(33) \qquad \theta\, \frac{\partial \mathcal{V}^*}{\partial \chi} + \vec{k}_J\, \frac{\partial \mathcal{F}_J(\mathcal{V}^*)}{\partial \chi} + c\left\{ \frac{\partial \mathcal{V}^*}{\partial t} + \frac{\partial \mathcal{F}_J(\mathcal{V}^*)}{\partial x_J} \right\} = 0 \, .$$

When expanded to order one in C, this equation may be written as

$$(34) \qquad \frac{\partial g_1^*}{\partial \chi} + \frac{\partial \mathcal{V}_0^*}{\partial t} + \frac{\partial \mathcal{F}_J(\mathcal{V}_0^*)}{\partial x_J} = 0 \, ,$$

where g_1 is linear with respect to $\mathcal{V}_1^*$. The basic requirement stated earlier leads to

$$(35) \qquad \frac{\partial \overline{\mathcal{V}_0^*}}{\partial t} + \frac{\partial \overline{\mathcal{F}_J(\mathcal{V}_0^*)}}{\partial x_J} = 0 \, .$$

Here we must stress a very simple but crucial argument : thanks to the conservative form of (31), the equation (33) holds throughout, even when $\mathcal{V}^*$ is discontinuous across the very many vortex sheets embedded in the flow. We leave the details to the reader.

The main difficulty with (35) is that it is a set of equations which do   not form a closed system of evolution equations for the five components of $\overline{\mathcal{V}_0^*}$, for the very simple reason that $\overline{\mathcal{F}_J(\mathcal{V}_0^*)}$ is not equal to $\mathcal{F}_J(\overline{\mathcal{V}_0^*})$. In the case of rolled vortex sheets $\mathcal{U}_0^*$ was independent of χ and the difficulty simply did not occur at all. Here the model is incomplete as long as we have not brought into it an algorithm for closing (35). We examine now this point.

Let us come back to (11), it shows that we may set

$$(36) \qquad \theta + \vec{k} \cdot \vec{u}^* = C\, w^* ,$$

in such a way that w^* remains bounded and, even, tends to a definite limit w_0^* when $C \to 0$. In order to accomplish our goal, we introduce some conventions regarding the notations. Let $\vec{F}^*$ be any vector, we decompose it according to its tangential and normal components with respect to $\chi = $ const, namely

$$(37) \qquad \vec{F}^* = \vec{F}_\tau^* + F_n^* \frac{\vec{k}}{|\vec{k}|} \qquad , \qquad \vec{k} \cdot \vec{F}_\tau^* = 0 ,$$

and, correspondingly

$$(38) \qquad \vec{u}^* = \vec{u}_\tau^* + v^* \vec{e} \qquad , \qquad \vec{k} = |\vec{k}|\, \vec{e} \qquad , \qquad \vec{u}_\tau^* \cdot \vec{e} = 0 .$$

If $(\vec{\cdot})$ stands for a vector valued algebraic quantity we will set accordingly $(\vec{\cdot})_\tau$ for its tangential component and $(\cdot)_n$ for its normal one. We shall complete later our notational conventions. Let us write the Euler equations as follows :

$$(39) \qquad \begin{cases} \rho^* \left(\frac{\partial \vec{u}^*}{\partial t} + \vec{u}^* \cdot \vec{\nabla} \vec{u}^* + w^* \frac{\partial \vec{u}^*}{\partial \chi} \right) + \frac{1}{\gamma M^2} \left(\vec{\nabla} p^* + C^{-1} \frac{\partial p^*}{\partial \chi} \vec{k} \right) = 0 , \\[2mm] \frac{\partial \rho^*}{\partial t} + \vec{u}^* \cdot \vec{\nabla} \rho^* + w^* \frac{\partial \rho^*}{\partial \chi} + \rho^* \left(\vec{\nabla} \cdot \vec{u}^* + \frac{\partial w^*}{\partial \chi} \right) = 0 , \\[2mm] \frac{\partial S^*}{\partial t} + \vec{u}^* \cdot \nabla S^* + w^* \frac{\partial S^*}{\partial \chi} = 0 . \end{cases}$$

If we consider $\vec{x}^*, \chi$ as four cartesian coordinates in a space of four dimensions and $\vec{u}^*, w^*$ as the corresponding velocity components, we may observe that (39) reminds us of somewhat boundary layer like equations, for a thin layer in the direction of χ , with viscous terms omitted. The analogy is not quite complete however. Although $\partial p_0^*/\partial \chi = 0$ is the exact analogue of the corresponding equation in boundary layer theory, we observe that p_0^* is not known. We must adopt our strategy to the problem at hand. Let us complete the notational convention by setting

$$(40) \qquad \begin{cases} \vec{\mathcal{M}}^* \equiv \rho^* \left(\frac{\partial \vec{u}^*}{\partial t} + \vec{u}^* \cdot \vec{\nabla} \vec{u}^* + w^* \frac{\partial \vec{u}^*}{\partial \chi} \right) + \frac{1}{\gamma M^2} \vec{\nabla} p^* , \\[2mm] \mathcal{R}^* \equiv \frac{\partial \rho^*}{\partial t} + \vec{u}^* \cdot \vec{\nabla} \rho^* + w^* \frac{\partial \rho^*}{\partial \chi} + \rho^* \left(\vec{\nabla} \cdot \vec{u}^* + \frac{\partial w^*}{\partial \chi} \right) , \\[2mm] \Sigma^* \equiv \frac{\partial S^*}{\partial t} + \vec{u}^* \cdot \vec{\nabla} S^* + w^* \frac{\partial S^*}{\partial \chi} , \end{cases}$$

where the sign $\equiv$ stands for an equality serving as a definition for the quantity on the left hand side. Expanding (39), with respect to C, we find

$$(41) \qquad \begin{cases} \vec{\mathcal{M}}_{0\tau}^* = 0 \quad , \quad \mathcal{M}_{0n}^* + \frac{1}{\gamma M^2} \frac{\partial \pi_1^*}{\partial \chi} = 0 , \\[2mm] \mathcal{R}_0^* = 0 \quad , \quad \Sigma_0^* = 0 . \end{cases}$$

To order zero let us count the number of unknown functions

$$(42)\quad\begin{cases}\vec{u}_{o\tau}^{x}=\overline{\vec{u}_{o\tau}^{x}}(t,\vec{x})+\widetilde{\vec{u}_{o\tau}^{x}}(t,\vec{x},\chi)\,,\;\cdots\;4\;\text{unknowns}\,,\\[4pt]w_{o}^{x}=\overline{w_{o}^{x}}(t,\vec{x})+\widetilde{w_{o}^{x}}(t,\vec{x},\chi)\,,\,\ldots\ldots\,2\quad''\quad,\\[4pt]v_{o}^{x}=\overline{v_{o}^{x}}(t,\vec{x})\,,\,\ldots\cdots\cdots\,1\quad''\quad,\\[4pt]\rlap{/}{t}_{o}^{x}=\overline{\rlap{/}{t}_{o}^{x}}(t,\vec{x})\,,\,\ldots\cdots\,1\quad''\quad,\\[4pt]\rho_{o}^{x}=\overline{\rho_{o}^{x}}(t,\vec{x})+\widetilde{\rho_{o}^{x}}(t,\vec{x},\chi)\,,\,\ldots\,2\quad''\quad,\\[4pt]S_{o}^{x}=\overline{S_{o}^{x}}(t,\vec{x})+\widetilde{S_{o}^{x}}(t,\vec{x},\chi)\,,\,\ldots\,2\quad''\quad,\\[4pt]\chi_{o}=\chi_{o}(t,\vec{x})\,,\,\ldots\cdots\,1\quad''\quad,\\[4pt]\vec{e}=\vec{e}(t,\vec{x})\,,\;|\vec{e}|=1\,\ldots\ldots\,2\quad''\quad,\end{cases}$$

so that we find a total of 15 unknown functions, out of which 10 are functions of $t,\vec{x}$ while 5 of them are also functions of χ . We have at our disposal the equations in (41), $\vec{u}_{o\tau}^{x}\cdot\vec{e}=0$ which is consequence of (38), the definition of $\vec{e}$ which appears also in (38) and (11). If we consider the second equation in (41), we observe that, by taking the mean, we eliminate the extra unknown $\rlap{/}{t}_{1}^{x}$. We list the equations

$$(43)\quad\widetilde{\vec{M}_{o\tau}^{x}}=0\,,\quad\widetilde{R_{o}^{x}}=0\,,\quad\widetilde{\Sigma_{o}^{x}}=0\,,\quad\widetilde{\rho_{o}^{x\gamma}e\,S_{o}^{x}}=0\,,\quad\widetilde{\vec{u}_{o\tau}^{x}}\cdot\vec{e}=0$$

$$(44)\quad\begin{cases}\overline{\vec{M}_{o\tau}^{x}}=0\,,\quad\overline{\vec{M}_{on}^{x}}=0\,,\quad\overline{\vec{u}_{o\tau}^{x}}\cdot\vec{e}=0\,,\\[4pt]\overline{R_{o}^{x}}=0\,,\quad\overline{\Sigma_{o}^{x}}=0\,,\quad\overline{\rlap{/}{t}_{o}^{x}}=\overline{\rho_{o}^{x\gamma}e\,S_{o}^{x}}\,,\\[4pt]\dfrac{\partial\chi_{o}}{\partial t}+\overline{v_{o}^{x}}\,|\vec{\nabla}\chi_{o}|=0\,,\\[4pt]\vec{e}=|\vec{\nabla}\chi_{o}|^{-1}\,\vec{\nabla}\chi_{o}\,;\end{cases}$$

there are 15 of them, just as many equations as there are unknowns. In order to investigate whether it is reasonable to consider that we have a closed system of equations, we must detail their structure . Corresponding to (43), we have

$$(45)\quad\begin{cases}\overline{\rho_{o}^{x}\Big(\dfrac{\partial u_{o\tau}^{x}}{\partial t}\Big)_{\tau}}+\overline{\rho_{o}^{x}\Big(\vec{u}_{o\tau}^{x}\cdot\vec{\nabla}\vec{u}_{o\tau}^{x}\Big)_{\tau}}+\overline{\rho_{o}^{x}w_{o}^{x}\dfrac{\partial\vec{u}_{o\tau}^{x}}{\partial\chi}}+\overline{\rho_{o}^{x}\,\rlap{/}{e}\{\vec{u}_{o\tau}^{x}\cdot\vec{\nabla}_{\tau}e+\vec{e}\cdot\vec{\nabla}\vec{u}_{o\tau}^{x}\}_{\tau}}+\dfrac{1}{\gamma M^{2}}\vec{\nabla}\rlap{/}{t}_{o}^{x}=0\,,\\[6pt]\overline{\dfrac{\partial S_{o}^{x}}{\partial t}+\vec{u}_{o\tau}^{x}\cdot\vec{\nabla}S_{o}^{x}+\vec{v}_{o}^{x}\vec{e}\cdot\vec{\nabla}S_{o}^{x}+w_{o}^{x}\dfrac{\partial S_{o}^{x}}{\partial\chi}}=0\,,\\[6pt]\overline{\dfrac{\partial w_{o}^{x}}{\partial\chi}+\vec{\nabla}\cdot\vec{u}_{o\tau}^{x}+\vec{e}\cdot(\vec{\nabla}\vec{u}_{o\tau}^{x})\cdot\vec{e}+\dfrac{1}{\rho_{o}^{x}}\dfrac{\partial\rho_{o}^{x}}{\partial t}+\dfrac{\vec{u}_{o\tau}^{x}\cdot\vec{\nabla}_{\tau}\rho_{o}^{x}}{\rho_{o}^{x}}+\vec{v}_{o}^{x}\dfrac{\vec{e}\cdot\vec{\nabla}\rho_{o}^{x}}{\rho_{o}^{x}}+\dfrac{w_{o}^{x}}{\rho_{o}^{x}}\dfrac{\partial\rho_{o}^{x}}{\partial\chi}}=0\,,\\[6pt]\overline{\rho_{o}^{x\gamma}e\,S_{o}^{x}}=0\,,\quad\overline{\vec{u}_{o\tau}^{x}}\cdot\vec{e}=0\,,\end{cases}$$

while, corresponding to (44) we find

$$\begin{cases}\dfrac{\partial\chi_{o}}{\partial t}+\vec{v}_{o}^{x}\cdot\vec{\nabla}\chi_{o}=0\quad,\quad\vec{e}=\dfrac{\vec{\nabla}\chi_{o}}{|\vec{\nabla}\chi_{o}|}\quad,\quad\vec{e}\cdot\overline{\vec{u}_{o\tau}^{x}}=0\\[8pt]\overline{\rlap{/}{t}_{o}^{x}}=\overline{\rho_{o}^{x\gamma}e\,S_{o}^{x}}\,,\end{cases}$$

$$(46)\begin{cases}
\left(\dfrac{\partial \vec{u}_{o_T}^{*}}{\partial t} + \overline{\vec{u}_{o_T}^{*}\cdot\vec{\nabla}_T\vec{u}_{o_T}^{*}} + \vec{V}_o^{*}\vec{e}\cdot\vec{\nabla}\vec{u}_{o_T}^{*} + \overline{w_o^{*}\dfrac{\partial \vec{u}_{o_T}^{*}}{\partial x}}\right)_T + \\[2mm]
\qquad + \vec{V}_o^{*}\left(\dfrac{\partial \vec{e}}{\partial t} + \vec{u}_{o_T}^{*}\cdot\vec{\nabla}\vec{e} + V_o^{*}\vec{e}\cdot\vec{\nabla}\vec{e}\right)_T + \dfrac{1}{\gamma M^2}\dfrac{1}{\overline{P}_o^{*}}\vec{\nabla}_T\overline{P}_o^{*} = 0\ , \\[4mm]
\overline{P}_o^{*}\dfrac{\partial\vec{V}_o^{*}}{\partial t} + \overline{P_o^{*}\vec{u}_{o_T}^{*}\cdot\vec{\nabla}_T V_o^{*}} + \overline{P_o^{*}V_o^{*}\vec{e}\cdot\vec{\nabla}v_o^{*}} + \\[2mm]
\qquad + \overline{P_o^{*}\left(\dfrac{\partial \vec{u}_{o_T}^{*}}{\partial t} + \vec{u}_{o_T}^{*}\cdot\vec{\nabla}_T\vec{u}_{o_T}^{*} + V_o^{*}\vec{e}\cdot\vec{\nabla}\vec{u}_{o_T}^{*}\right)}\cdot\vec{e} = 0\ , \\[4mm]
\overline{\dfrac{\partial S_o^{*}}{\partial t}} + \overline{u_{o_T}^{*}\cdot\nabla S_o^{*}} + \overline{V_o^{*}\vec{e}\cdot\vec{\nabla}S_o^{*}} + \overline{w_o^{*}\dfrac{\partial S_o^{*}}{\partial x}} = 0\ , \\[4mm]
\overline{\dfrac{\partial P_o^{*}}{\partial t}} + \overline{u_{o_T}^{*}\cdot\nabla_T P_o^{*}} + \overline{V_o^{*}\vec{e}\cdot\nabla P_o^{*}} + \overline{P_o^{*}V_o^{*}\vec{\nabla}_T\cdot\vec{e}} + \\[2mm]
\qquad + \overline{P_o^{*}\vec{\nabla}_T\cdot\vec{u}_{o_T}^{*}} + \overline{P_o^{*}\vec{e}\cdot(\vec{\nabla}_T\vec{u}_{o_T}^{*})\cdot\vec{e}} = 0\ .
\end{cases}$$

These are a quite intricate set of equations. The only point we want to investigate is whether they form a closed set of evolution equations for the 15 unknowns (42). Assume that each of these is known as a function of $\vec{X}$ and, when appropriate, of χ , then we want to compute them, in principle at least, at $t+\Delta t$ using (45) (46). For our purpose, we may rewrite (46) as follows

$$(47)\begin{cases}
\left(\dfrac{\partial\vec{u}_{o_T}^{*}}{\partial t}\right)_T + \vec{V}_o^{*}\dfrac{\partial\vec{e}}{\partial t} = \vec{\gamma}_T\ , \quad \dfrac{\partial\vec{V}_o^{*}}{\partial t} = \vec{\gamma}_n\ , \quad \dfrac{\partial S_o^{*}}{\partial t} = \sigma\ , \quad \dfrac{\partial P_o^{*}}{\partial t} = m\ , \\[3mm]
\dfrac{\partial \chi_o}{\partial t} = -V_o^{*}|\nabla\chi_o|\ ,
\end{cases}$$

$$(48)\qquad \vec{e} = |\vec{\nabla}\chi_o|^{-1}\vec{\nabla}\chi_o\ , \quad \vec{e}\cdot\vec{u}_{T_o}^{*} = 0\ ,$$

$$(49)\qquad \overline{P}_o^{*} = \overline{P_o^{*\gamma}e^{S_o^{*}}}\ .$$

We observe that $(\partial\vec{u}_{o_T}^{*}/\partial t)_T$ is only part of $\partial\vec{u}_{o_T}^{*}/\partial t$ but the subsidiary information is provided by the second of (48) [where $\vec{e}$ is evaluated at $t+\Delta t$ from the first of (48) and from the last of (47)]. As a consequence we may state that (47) (48) allow to compute $\vec{u}_{o_T}^{*}$, V_o^{*} , $\overline{P}_o^{*}$, χ_o , $\vec{e}$ at time $t+\Delta t$ as functions of χ . Let us write now (45) as follows

$$(50)\begin{cases}
\widetilde{P_o^{*}\left(\dfrac{\partial\vec{u}_{o_T}^{*}}{\partial t}\right)_T} = \widetilde{\vec{\gamma}}_T\ , \quad \dfrac{\partial\widetilde{S_o^{*}}}{\partial t} = \widetilde{\sigma}\ , \quad \widetilde{\vec{u}_{o_T}^{*}}\cdot\vec{e} = 0\ , \\[3mm]
\widetilde{P_o^{*\gamma}e^{S_o^{*}}} = 0\ ,
\end{cases}$$

$$(51)\qquad \dfrac{\partial\widetilde{w_o^{*}}}{\partial x} = \widetilde{\phi} - \dfrac{1}{P_o^{*}}\widetilde{\dfrac{\partial P_o^{*}}{\partial t}} = \widetilde{\phi} + \dfrac{1}{\gamma}\dfrac{\partial\widetilde{S_o^{*}}}{\partial t}\ .$$

From the second of (50), we get $\partial \widetilde{S}_o^x / \partial t$ which allows to compute $\widetilde{S}_o^x$ at $t+\Delta t$; as we know already $\overline{S}_o^x$ we have a full knowledge of S_o^x at $t+\Delta t$. From $\overline{\rho_o^x \gamma e^{S_o^x}} = 0$, the knowledge of S_o^x and the one of $\overline{\rho_o^x}$ at $t+\Delta t$ we achieve the one of ρ_o^x at $t+\Delta t$; then (49) gives $\rho_o^x = \overline{\rho_o^x}$, at $t+\Delta t$. From the first and the third of (50) we may compute $\partial \overline{u}_{o_T}^x / \partial t$ thanks to our previous knowledge of $\partial \overline{u}_{o_T}^x / \partial t$. We achieve the cycle by computing $\widetilde{w}^x$ from (51), but there is a subtle point about this last part of the cycle because, as (51) shows we ought to know $\partial \overline{S}_o^x / \partial t$ at $t+\Delta t$. From the second of (45) we may rewrite (51) as follows

$$(52) \qquad \frac{\partial \widetilde{w}_o^x}{\partial X} + \frac{\overline{w_o^x}}{\gamma} \frac{\partial \overline{S}_o^x}{\partial X} = \widetilde{\psi} \, ,$$

where $\widetilde{\psi}$ and $\partial \overline{S}_o^x / \partial X$ may be considered as known at $t+\Delta t$. Using once more $\overline{\rho_o^x \gamma e^{S_o^x}} = 0$, we may replace (52) by

$$(53) \qquad \rho_o^x \frac{\partial}{\partial X} \left(\frac{w_o^x}{\rho_o^x} \right) = \widetilde{\psi} \, ,$$

and, from this, it is obvious that $\widetilde{w}_o^x$ at $t+\Delta t$ can be computed up to an arbitrary function of t and $\vec{x}$. We may make w_o^x completely defined by observing that, whenever there are vortex sheets embedded into the fine scale structure, we may enforce that χ is constant on them, which requires that w_o^x be zero on them.

The previous analysis gives some confidence about the mathematical conjecture that (43)(44) is a closed system for the unknowns (42). Of course, we do not know any numerical code which would render the numerical solution more practical than brute force that would consist in using a standard three dimensional code, with a mesh sufficiently refined, in order to be able to capture the fine scale structure of the flow. However, it is appropriate here to make the following observation . The suitableness of any numerical code for computing the flow would rely on the fact that the mesh be refined only where this is strictly necessary. Now a moment's reflexion shows that this is just what has been achieved through asymptotic modelling. As a matter of fact let $\Delta \vec{x}$ be the spatial [three-dimensional] mesh, and $\Delta \chi$ be the mesh size corresponding to χ . What we have done is to use a refined spatial mesh in the direction normal to $\chi = $ const and a normal mesh for other directions. It is even almost obvious that this has been done in an optimum way. Further research is needed in order to explore this avenue; we leave it, and explore another direction. We assume from now on that the fluctuations are small in some sense. In order to render this precise, we introduce a small parameter ε and assume that we may use an expansion

$$(54) \quad \left\{ \begin{array}{l} W_o^x = \left(\vec{u}_{o_T}^x, w_o^x, \rho_o^x, \rho_o^x, S_o^x, \theta_o, k_o, \chi_o \right)^T, \\[4pt] W_o^x = W_{oo}^x + \varepsilon\, W_{o_1}^x + O(\varepsilon^2), \\[4pt] W_{oo}^x = W_{oo}^x(t, \vec{x}) , \\[4pt] W_{o_1}^x = \overline{W_{o_1}^x}(t, \vec{x}) + \widetilde{W_{o_1}^x}(t, x, \chi) , \end{array} \right.$$

where we have, obviously :

$$(55) \qquad W_{oo}^x = \left(\vec{u}_{oo}^x, 0, \rho_{oo}^x, \rho_{oo}^x, S_{oo}^x, \theta_o, \vec{k}_o, \chi_o \right)^T,$$

$$(56) \qquad \widetilde{W}_{o_1}^x = \left(\widetilde{\vec{u}}_{o_1}^x, \widetilde{W}_{o_1}^x, 0, \widetilde{\rho}_{o_1}^x, \widetilde{S}_{o_1}^x, 0, 0, 0 \right)^T.$$

Let us come back to (35) and make the observation that

$$(57) \quad \widetilde{F}_J(\mathcal{V}_{oo}^x + \varepsilon \mathcal{V}_{o_1}^x + \varepsilon^2 \mathcal{V}_{o_2}^x) = F_J(\mathcal{V}_{oo}^x) + \varepsilon\, G_J(\mathcal{V}_{oo}^x, \mathcal{V}_{o_1}^x) + \varepsilon^2 \left\{ G_J(\mathcal{V}_{oo}^x, \mathcal{V}_{o_2}^x) + K_J(\mathcal{V}_{oo}^x, \mathcal{V}_{o_1}^x) \right\} + \cdots$$

with the obvious remark that G_J is linear with respect to $\overline{V_{o1}^x}$. We find, obviously

$$(58) \quad \overline{F_J(V_o^x)} = F_J(\overline{V_{oo}^x}) + \varepsilon\, G_J(\overline{V_{oo}^x}, \overline{V_{o1}^x}) + \varepsilon^2\left\{ G_J(\overline{V_{oo}^x}, \overline{V_{o2}^x}) + \overline{K_J(V_{oo}^x, V_{o1}^x)} \right\} + \cdots$$

and (35), when expanded, leads to a whole hierarchy of equations

$$(59) \quad \begin{cases} \dfrac{\partial \overline{V_{oo}^x}}{\partial t} + \dfrac{\partial\, F_J(\overline{V_{oo}^x})}{\partial x_J} = 0, \\[2mm] \dfrac{\partial \overline{V_{o1}^x}}{\partial t} + \dfrac{\partial}{\partial x_J}\left\{ G_J(\overline{V_{oo}^x}, \overline{V_{o1}^x}) \right\} = 0, \\[2mm] \dfrac{\partial \overline{V_{o2}^x}}{\partial t} + \dfrac{\partial}{\partial x_J}\left\{ G_J(\overline{V_{oo}^x}, \overline{V_{o2}^x}) + \overline{K_J(V_{oo}^x, V_{o1}^x)} \right\} = 0. \end{cases}$$

We have the obvious but quite illuminating result that, up to order ε , the mean values may be computed at the outset even without any knowledge of the fluctuations. Let us examine the equation for $\overline{V_{o2}^x}$; it looks like the one for $\overline{V_{o1}^x}$ with a right hand side. We observe that K_J is quadratic with respect to V_{o1}^x , namely

$$(60) \quad \begin{cases} K_J(V_{oo}^x, \lambda\, V_{o1}^x) = \lambda^2\, K_J(V_{oo}^x, V_{o1}^x), \\[2mm] K_J(V_{oo}^x, V_{o1}^{x\,\prime} + V_{o1}^{x\,\prime\prime}) = K_J(V_{oo}^x, V_{o1}^{x\,\prime}) + \hat{K}_J(V_{oo}^x; V_{o1}^{x\,\prime}, V_{o1}^{x\,\prime\prime}) + K_J(V_{oo}^x, V_{oo}^{x\,\prime\prime}), \end{cases}$$

where $\hat{K}_J$ is bilinear, namely

$$(61) \quad \hat{K}_J(V_{oo}^x; \lambda A, \mu B) = \lambda \mu\, \hat{K}_J(V_{oo}^x; A, B).$$

Thanks to the above, we may compute

$$(62) \quad \overline{K_J(V_{oo}^x, V_{o1}^x)} = K_J(V_{oo}^x, \overline{V_{o1}^x}) + \overline{K_J(V_{oo}^x, \tilde{V}_{o1}^x)},$$

where we have used the obvious relation

$$(63) \quad \overline{\hat{K}_J(V_{oo}^x; \overline{V_{o1}^x}, \tilde{V}_{o1}^x)} = 0.$$

As a consequence, provided we are able to compute the fine scale structure $\tilde{W}_{o1}^x$ to lower order, we may compute the mean flow up to order ε^2 through use of a slight variant of one and the same numerical code capable of handling the Euler equations. We may be even more explicit than that. Let us start from

$$(64) \quad V_{oo}^x + \varepsilon\, \overline{V_{o1}^x} + \varepsilon^2\, \overline{V_{o2}^x} = \hat{V}_o^x,$$

we get

$$(65) \quad \begin{aligned} F_J(\hat{V}_o^x) &= F_J(V_{oo}^x) + \varepsilon\, G_J(V_{oo}^x, \overline{V_{o1}^x}) + \\ &\quad + \varepsilon^2\left\{ G_J(V_{oo}^x, \overline{V_{o2}^x}) + K_J(V_{oo}^x, \overline{V_{o1}^x}) \right\} + O(\varepsilon^3), \end{aligned}$$

and, from (59), we find

$$(66) \quad \frac{\partial \tilde{\mathcal{P}}_0^{*}}{\partial t} + \frac{\partial}{\partial x_J}\left(\mathcal{F}_J(\tilde{\mathcal{P}}_0^{*})\right) + \frac{\partial}{\partial x_J}\left(\overline{K_J(\mathcal{V}_{00}^{*}, \tilde{\mathcal{P}}_{01}^{*})}\right) = O(\varepsilon^3).$$

Now, we may state that, once $\mathcal{V}_{00}^{*}$ and $\tilde{\mathcal{P}}_{01}^{*}$ are known, we may compute $\hat{\mathcal{P}}_0^{*}$, that is $\mathcal{V}_0^{*}$ up to $O(\varepsilon^2)$, by using, only once, a standard code for the Euler equations completed by another one for computing the non homogeneous term $K_J(\mathcal{V}_{00}^{*}, \tilde{\mathcal{P}}_{01}^{*})$. We conjecture that, thanks to the linearity with respect to $\tilde{\mathcal{P}}_{01}^{*}$ in the second of (59), there exists a number of circumstances for which $\tilde{\mathcal{P}}_{01}^{*} = 0$. Each time this holds we may get an accuracy $O(\varepsilon^2)$ in the means by using a three step procedure : i) compute $\mathcal{V}_{00}^{*}$ from an Euler code, ii) compute $\tilde{\mathcal{P}}_{01}^{*}$ from a code to be considered from now on iii) compute $\tilde{\mathcal{P}}_0^{*}$ by using the variant of the Euler code including the non homogeneous term $K_J(\mathcal{V}_{00}^{*}, \tilde{\mathcal{P}}_{01}^{*})$ which may be computed through quadrature in one dimension. Let us now consider how $\tilde{\mathcal{P}}_{01}^{*}$ may be computed. Rather than using (45) we prefer to go back to the vorticity and entropy equations. The second one appears already as the third of (39), while the vorticity equation may be written, using (28), as follows

$$
\begin{aligned}
(67) \quad &\frac{\partial \vec{\Omega}^{*}}{\partial t} + \vec{u}^{*}\cdot\vec{\nabla}\vec{\Omega}^{*} + w^{*}\frac{\partial \vec{\Omega}^{*}}{\partial \chi} + \left(\vec{\nabla}\cdot\vec{u}^{*} + \frac{\partial w^{*}}{\partial \chi}\right)\vec{\Omega}^{*} - \vec{\Omega}^{*}\cdot\vec{\nabla}\vec{u}^{*} - \\
&- \vec{k}\cdot(\vec{\nabla}_{\wedge}\vec{u}^{*})\frac{\partial \vec{u}^{*}}{\partial \chi} - \frac{1}{\gamma M^2}\frac{1}{\rho^{*2}}\left(\frac{\partial k^{*}}{\partial \chi}\vec{\nabla}\rho^{*}_{\wedge}\vec{k} + \frac{\partial \rho^{*}}{\partial t}\vec{k}_{\wedge}\vec{\nabla}\rho^{*}\right) - \\
&- C\frac{1}{\gamma M^2}\frac{1}{\rho^{*2}}\vec{\nabla}\rho^{*}_{\wedge}\vec{\nabla}\rho^{*} = 0,
\end{aligned}
$$

and we find

$$
(68) \quad
\begin{cases}
\dfrac{\partial \vec{\Omega}_{01}^{*}}{\partial t} + \vec{u}_{00}^{*}\cdot\vec{\nabla}\vec{\Omega}_{01}^{*} + (\vec{\nabla}\cdot\vec{u}_{00}^{*})\vec{\Omega}_{01}^{*} - \vec{k}_0\cdot(\vec{\nabla}_{\wedge}\vec{u}_{00}^{*})\dfrac{\partial \vec{u}_{01}^{*}}{\partial \chi} - \tilde{\vec{\Omega}}_{01}^{*}\cdot\vec{\nabla}\vec{u}_{00}^{*} - \\
\qquad\qquad - \dfrac{1}{\gamma M^2}\dfrac{1}{\rho_0^{*2}}\vec{k}_0_{\wedge}\vec{\nabla}\rho_0^{*}\dfrac{\partial \tilde{\rho}_{01}^{*}}{\partial \chi} = 0, \\[2mm]
\dfrac{\partial \tilde{S}_{01}^{*}}{\partial t} + \vec{u}_{00}^{*}\cdot\vec{\nabla}\tilde{S}_{01}^{*} + \vec{u}_{01}^{*}\cdot\vec{\nabla}\tilde{S}_{00}^{*} = 0,
\end{cases}
$$

to which must be added

$$(69) \quad \vec{\Omega}_{01}^{*} = \vec{k}_0_{\wedge}\frac{\partial \vec{u}_{01}^{*}}{\partial \chi} \quad, \quad \rho_{00}^{*}\tilde{S}_{01}^{*} + \gamma \tilde{\rho}_{01}^{*} = 0,$$

according to (29), and to the fourth of (50) and (54). The final result is that $\vec{\Omega}_{01}^{*}$ and $\partial \tilde{S}_{01}^{*}/\partial \chi$ may be substituted respectively to $\vec{V}$ and Σ into (16). Let $\mathcal{F}_{00}$ be the analogue of the operator $\mathcal{F}_0$ computed from the field $\vec{u}_{00}^{*}, k_0^{*}, S_{00}^{*}$, then we have

$$(70) \quad \left(\vec{k}_0_{\wedge}\vec{u}_{01}^{*}, \tilde{S}_{01}^{*}\right) = \mathcal{F}_{00}(t^0,t)\left(\vec{k}_0_{\wedge}\vec{u}_{01}^{*0}, \tilde{S}_{01}^{*0}\right).$$

We may conclude from the previous analysis that the case when there is a fine scale structure to the order $O(1)$, provided that the corresponding fluctuations be small [of order $O(\varepsilon)$] is no more complex than the one when such a fine scale structure occurs only at order $O(C)$. If we are interested only in the effect of the fine scale structure on the mean flow, this effect may be computed up to $O(\varepsilon^2)$ through use of a standard numerical code for the Euler equations supplemented by a code for computing the transport operator $\mathcal{F}_{00}$, and another one for effecting the quadrature involved in $K_J(\mathcal{V}_{00}^{*}, \tilde{\mathcal{P}}_{01}^{*})$.

4. Conclusion

We briefly report on the last topic concerning highly swirling and/or
slender core flows. The starting point consists in Navier–Stokes equations
written in non dimensional polar coordinates r, θ, z with D and L as units of
lengths in the respective radial and axial directions. The corresponding components of velocity are u, v, w with $DL^{-1}W, V$ and W as units, while the pressure variation is $\rho V^2 p$, and time is $T t$. We find four dimensionless parameters within the set of equations. They are a Strouhal number $S = L/WT$, a
slenderness ratio $\alpha = D/L$, an inverse swirling parameter $\mathcal{E} = W/V$ and a
radial Reynolds number $N = \alpha^2 W L/\nu$, where ν is the kinematic viscosity. We
observe that $\alpha = (N/R_e)^{1/2}$ where $R_e = WL/\nu$ is a more conventional Reynolds
number. The situations corresponding to high swirl $(\mathcal{E} \ll 1)$ and/or slender cores
$(\alpha \ll 1)$ is relevant to a number of industrial (see Lewellen [1971]) or geophysical (see Morton [1966]) flow configurations. We have investigated a
number of limiting processes in which S and N remain 0 (1) with $\alpha \to 0$ while
$\mathcal{E}$ is either 0 (1) or tends to zero or infinity. Here, we report briefly on
$\mathcal{E} = 1, \quad \alpha \to 0$. Setting $\mathcal{U} = (u, v, w, p)^T$ we expand according to

$$(1) \qquad \mathcal{U} = \mathcal{U}_0 + \alpha\, \mathcal{U}_1 + \cdots \ .$$

The most prominent fact is that, to zeroth order, only $\mathcal{U}_0$, the radial
component of velocity, may depend on θ , while p_0 and V_0 are related by the
classical radial equilibrium equation. If we use the notation $\overline{\mathcal{U}}_0$ for the
azimuthal average of $\mathcal{U}_0$, we find, to the next order, a closed set of equations governing $\overline{\mathcal{U}}_0, W_0, p_0, V_0$, and this set is the one which rules axisymmetric flow with swirl. This is well known. What is perhaps less known is
that, if we set $\mathcal{U}_0 = \overline{\mathcal{U}}_0 + \widetilde{\mathcal{U}}_0$, so that the average of $\widetilde{\mathcal{U}}_0$ over θ is zero,
then we may find a linear system for $\widetilde{u}_0, \widetilde{v}_1, \widetilde{p}_1$ and a study of this system
reveals that the way in which $\widetilde{\mathcal{U}}_0$ may depend on θ is rather restricted.
Assume that the flow is unconfined radially and that there exists some cons-
tant n_0 such that

$$(2) \qquad n_0^2 - 2 + \frac{r}{V_0}\frac{\partial V_0}{\partial r} + \frac{r^2}{V_0}\frac{\partial^2 V_0}{\partial r^2} > 0 \ ,$$

then $\widetilde{\mathcal{U}}_0$ cannot have Fourier components $\cos(n\theta), \sin(n\theta)$ with n greater than n_0 .
If V_0 is a so-called viscous vortex, then one may set $n_0 > 1.9$ so that only
$\cos\theta$ and $\sin\theta$ are allowed. The same situation concerning $\widetilde{\mathcal{U}}_0$ occurs in a
number of other limiting processes. This reveals that, to leading order, highly
swirling flows that are unconfined radially have a very high tendency towards
axisymmetry.

We stop here this discussion of vortex flows. We hope that the point of
view emphasized in the introduction has been illustrated and that some readers
will be convinced that asymptotic techniques provide very powerful tools for
the purpose of modelling problems which are too stiff for pure numerical simulation. We are convinced that, the more computing will be efficient, the
more will be the need for techniques capable of unravelling stiff problems,
and asymptotic modelling, among others, proves to be an efficient tool
towards this purpose.

REFERENCES

K.C. ADAMSON, A.F. MESSITER [1980]. Ann. Rev. Fluid Mech. 12, p. 103-138.

G.K. BATCHELOR [1967]. Cambridge University Press.

S.N. BROWN [1965]. J.F.M. 22, p. 17-32.

S.N. BROWN, K.W. MANGLER [1967]. Aeron. Quart. 18, p. 354-366.

A.J. CALLEGARI, L. TING [1978]. SIAM. J. Appl. Math. 35, p. 148-176.

L.E. FRAENKEL [1970]. Proc. Roy. Soc. A 316, p. 29-62.

L.E. FRAENKEL [1972]. J.F.M. 51, p. 119-135.

J.P. GUIRAUD [1977]. p. 244-259. In C. M. BRAUNER, G. GAY, J. MATHIEU, Singular perturbations and boundary layer theory. Lecture notes in mathematics Vol. 594, SPRINGER, 1977.

J.P. GUIRAUD, R. ZEYTOUNIAN [1977]. J.F.M. 79, p. 93-102.

J.P. GUIRAUD, R. ZEYTOUNIAN [1978]. Int. Jal. Eng. Sci. 13, p. 515-526.

J.P. GUIRAUD, R. ZEYTOUNIAN [1979 a]. J.F.M. 101, p. 393-401.

J.P. GUIRAUD, R. ZEYTOUNIAN [1979 b]. J.F.M. 90, p. 197-201.

M.G. HALL [1961]. J.F.M. 11, p. 209-228.

H. HELMHOLTZ [1858]. Jal. Reine Angew. Math. 55, p. 25-65 [translated by P.G. TAIT. Phil. Mag. 23, p. 485-512, 1867].

W.M. HICKS [1884]. Phil. Trans. Roy. Soc. A 175, p. 183.

W.M. HICKS [1885]. Phil. Trans. Roy. Soc. A 176, p. 756.

S. HUBERSON [1980]. La Recherche Aérospatiale N° 1980-3, p. 197-204.

H. KADEN [1931]. Ing. Archive 2, p. 140.

S. KAPLUN [1954]. ZAMP 5, p. 111-136.

S. KAPLUN [1957]. Jal. Math. Mech. 6, p. 595-603.

S. KAPLUN, P.A. LAGERSTROM [1957]. Jal. Math. Mech. 6, p. 585-593.

D. KUCHEMAN [1965]. Progress in Aeronautical Sciences. Vol. 7. Pergamon Press.

M.T. LANDAHL [1981]. ZAMM 61, p. T9-T14.

J.C. LE BALLEUR [1981]. La Recherche Aérospatiale 1981 N°3, p. 161-185.

A. LEONARD [1980]. Jal Comp. Physics 37, p. 289-335.

W.S. LEWELLEN [1962]. J.F.M. 14, p. 420-432.

W.S. LEWELLEN [1964]. AIAA Jal. 3, p. 91.

W.S. LEWELLEN [1971]. NASA CR-1772.

K.W. MANGLER, J. WEBER [1967]. J.F.M. $\underline{30}$, p. 177-196.

D.W. MOORE [1975]. Proc. Roy. Soc. A $\underline{345}$, p. 417.

D.W. MOORE [1978]. Studies in Applied Math. $\underline{48}$, p. 119-140.

D.W. MOORE, P.G. SAFFMAN [1972]. Phil. Trans. Roy. Soc. A $\underline{276}$, p. 403-429.

B.R. MORTON [1966]. p. 145-194. In D. KUCHEMAN [1965].

V.Ya. NEILAND [1969]. Fluid Dynamics 1969, p. 33-35.

J.H. OLSEN, A. GOLDGURG, M. ROGERS [1971]. Aircraft wake turbulence and its protection. Plenum Press.

I. PROUDMAN, J.R.A. PEARSON [1957]. J.F.M. $\underline{2}$, p. 237-262.

C. REHBACH [1978]. In High angle of attack aerodynamics. AGARD Proceedings CP 247, 14-1 to 14-9.

P.G. SAFFMAN [1970]. Studies in Applied Mathematics $\underline{49}$, p. 371-380.

P.G. SAFFMAN, G.R. BAKER [1979]. Ann. Rev. Fluid Mech. $\underline{11}$, p. 95-122.

W.A. THOMSON [1867]. Phil. Mag. $\underline{34}$, p. 15.

J.J. THOMSON [1883]. A tratise on the motion of vortex rings. London McMillan.

L. TING, C. TUNG [1965]. Phys. Fluids $\underline{8}$, p. 1039-1051.

C. TUNG, L. TING [1967]. Phys. Fluids $\underline{10}$, p. 901-910.

S.E. WIDNALL, D. BLISS, A. ZALAY [1971], p. 305-338. In OLSEN and al. [1971].

H. LAMB, Hydrodynamics. Cambridge Univ. Press 1932.

K. STEWARTSON, P.G. WILLIAMS. Proc. Roy. Soc. $\underline{321}$, p. 181-206, [1969].

WAVE PROPAGATION, INSTABILITY, AND BREAKDOWN OF VORTICES

by

S. Leibovich

Sibley School of Mechanical and Aerospace Engineering
Cornell University
Ithaca, NY 14853 U.S.A.

Abstract

The themes of wave propagation and hydrodynamic stability have played prominent roles in attempts to understand the phenomenon of vortex break-down. Recent theoretical and experimental work provide evidence that waves and instabilities are important elements of the vortex breakdown process. Breakdown of vortex cores at high Reynolds numbers may occur in one of two forms, the apparently axisymmetric ("bubble") or the "spiral" form. The application of soliton theory and hydrodynamic stability theory to both forms is discussed, together with a new large amplitude theory for axisymmetric waves.

Introduction

The term "vortex breakdown" generally refers to a structural change in concentrated vortices with axial streaming embedded in an irrotational flow. The vortices in which this phenomenon has been observed are filaments whose centerlines have small curvatures compared to the curvatures of their cores, in which all of the vorticity is to be assumed confined. There are many kinds of structural changes that such vortices can undergo, and "vortex breakdown" most commonly refers to changes involving a deceleration of the axial velocity component occurring in an axial distance on the order of the vortex core diameter accompanied by a change in sign of the azimuthal vorticity and leading to (in some frame of reference) a reversal of flow on the vortex axis. (See Figure 1.) Since the variations of the properties of

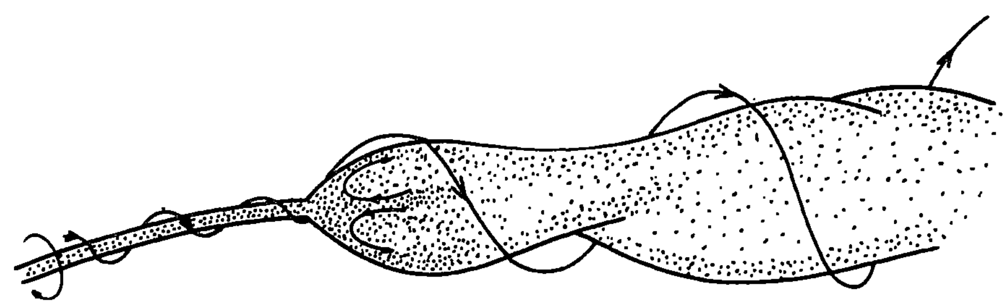

Figure 1.   Illustration of breakdown of a vortex filament.

the vortex (core radius and vorticity distribution) generally take place on a length scale large compared to the core diameter, vortex breakdown appears to be a sudden transformation of the vortex resulting in an almost step increase in core diameter and substantial pressure variations in the immediate vicinity of the breakdown. The large accelerations associated with the phenomenon suggests that it is inertially controlled with the influences of viscous diffusion being secondary. Experiments (see [1] for a review) tend to substantiate this view, since breakdown occurs at a core Reynolds number in the thousands, and exhibits no qualitative change in form as the Reynolds number is varied.

Figure 2 is a photograph of vortex breakdown in a water flowing from left to right in a tube. The photograph shows the two most common forms of breakdown, the "axisymmetric" (or bubble) and spiral forms, appearing simultaneously. The bubble is upstream of the spiral in this flow, and the stagnation points marking the characteristic flow reversals are indicated by arrows. The spiral form can occur independently and the bubble form often terminates in turbulence, so that the reformation of the flow in the wake of the bubble that, in the case illustrated, leads to the succeeding spiral breakdown does not generally occur. Although the photograph in Figure 2 does not show it, flow inside the bubble form is <u>not</u> axially symmetric, and azimuthal asymmetry seems to be characteristic of the bubble as well as the spiral form of breakdown.

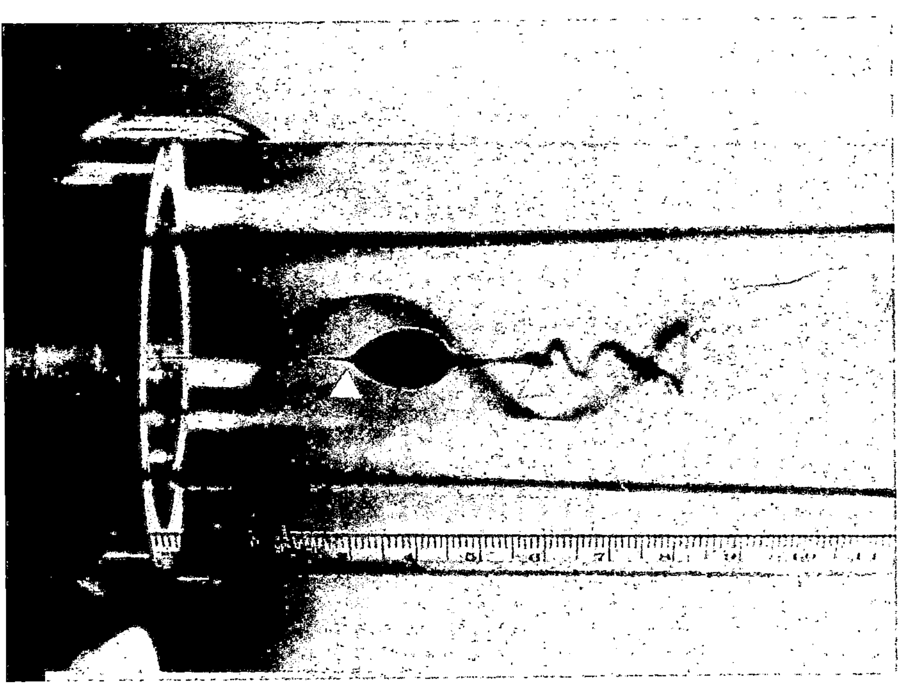

Figure 2. Photograph showing the simultaneous appearance of the "bubble" and the "spiral" forms of vortex breakdown.

Breakdown, as I have outlined it here, is an essentially nonlinear event
- there can be no weak vortex breakdowns. This is why identification of the
underlying physical mechanisms is so difficult. Theoretical investigations
nearly always depend upon linearization, and consequently can only be sugges-
tive. Existing theoretical work on vortex breakdown deals mostly with
axisymmetric events (although, as emphasized in [1], this may not be
appropriate), and analogies have been drawn to boundary layer separation and
to wave propagation phenomena. Ludwieg [2] pointed out (and Hall [3] further
explained) that the boundary layer separation criterion for the onset of
breakdown is equivalent to the criterion coming from wave theory - thus the
viscous force is not the agent directly responsible for breakdown.

Developments in the subject through 1978 have been reviewed by Leibovich
[1] and a earlier by Hall [3]. The present paper will focus on more recent
theoretical efforts related to wave propagation and instability phenomena
that occur in long slender vortex tubes, and that may play a role in vortex
breakdown. As in all previous theoretical work, the curvature of the vortex
axis is neglected.

The Existence of Axisymmetric Bubbles

The weight of the evidence strongly indicates that axially symmetric
bubbles, that is regions of closed streamsurfaces embedded in a streaming
vortex, are possible solutions of the full Navier-Stokes equations. Numerical solutions of the steady, axisymmetric Navier-Stokes equations containing
regions of flow reversal have been obtained by Kopecky and Torrance [4] and
by Grabowski and Berger [5]. Although questions can be raised concerning the
details of these calculations, there is little reason to doubt that they constitute a demonstration of the existence of solutions of the Navier-Stokes
equations of this type. In fact, such a flow has been experimentally created
by Vogel [6] and by Ronnenberg [7] by rotating the lid in a closed
container. Numerical solutions of the steady, axisymmetric Navier-Stokes
equations simulating Vogel's configuration have been carried out by Lugt and
Haussling [8], and they successfully predicted some aspects of the
experiment. The relationship between Vogel's bubble, and vortex breakdown in
concentrated vortices is unclear, however, and until the matter is better
understood, I prefer to reserve the term "vortex breakdown" for phenomena
occurring in concentrated vortices.

Laboratory experiments on concentrated vortices [9 - 12] show that
axisymmetric flows lose stability to asymmetric perturbations at Reynolds
numbers lower than the numerical experiments require for bubble formation.
Thus, while the axially symmetric solution may prove to play an important
role in the occurrence of the bubble form of breakdown, one may expect it to
be accompanied by asymmetric flow features as well. If we adopt a
cylindrical (r,θ,z) coordinate system, and think of a Fourier decomposition
of the flowfield in the azimuth θ with associated (integral) wavenumber m,
then each flow variable can be represented in a series in the form

$$\sum_{m=-\infty}^{\infty} C_m(r,z,t)e^{im\theta} .$$

The axisymmetric mode, $m=0$ is certainly important in the bubble form of
breakdown, but whether it dominates the nonaxisymmetric terms and is
therefore "close" to the strictly axisymmetric solutions computed to date
remains to be seen.

Inviscid Perturbations of Columnar Vortices

If the basic flow varies slowly in the axial direction, then the vortex
acts as a waveguide with slowly varying propagation or stability
characteristics. It is then possible [13] for waves to be trapped in a
definite axial interval (see Figure 3). This interval may be far removed
from the source of wave energy, which might originate, for example, from
disturbances created at the trailing edge of a wing or at the exit of a tube,
or from a region of instability some distance downstream. If the properties
of the vortex (core diameter, velocity distribution) vary slowly along its
length with a characteristic length ℓ, then weak perturbations on a length
scale much shorter than ℓ can be treated by supposing the basic motion in the
core is columnar, with velocity vector and pressure depending only upon the
radial distance r from the symmetry axis. Perturbations of this basic state
in the form of wavy motions are sought; the waves may be ordinary
propagating waves, or, if the basic states prove to be unstable, amplifying
normal modes.

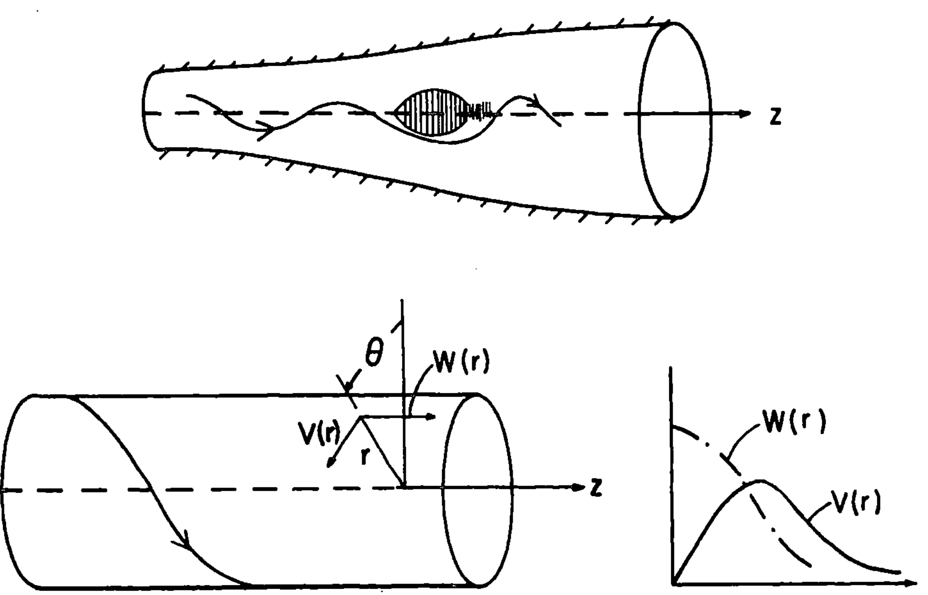

Figure 3. Top. Swirling motion in a slowly varying waveguide indicating
possible wave trapping.

Bottom. Columnar vortex idealization, showing coordinate systems
used in the analysis.

We therefore postulate the existence of a basic columnar flow with
pressure P(r), and velocity $(0,V(r),W(r))$ referred to a cylindrical (r,θ,z)
coordinate system and we write the perturbed variables as

$$\underset{\sim}{v} = \underset{\sim}{U} + \varepsilon\underset{\sim}{u} = \left(\varepsilon u, V+\varepsilon v, W+\varepsilon w\right), \quad p = P + \varepsilon\pi \tag{1}$$

where ε is an amplitude parameter assumed to be small. We assume that all
distances have been made dimensionless by reference to a characteristic
radial distance R, that velocities have been referred to a characteristic

speed (say V_0) in the basic flow, and that time has been scaled with the convective time R/V_0.

The dependent variables (u,v,w,π) satisfy the following equations

$$\underset{\sim}{u}_t + \underset{\sim}{U} \cdot \hat{\nabla}\underset{\sim}{u} + C\underset{\sim}{u} + \nabla\pi = \varepsilon\underset{\sim}{N} \qquad \text{(a)}$$

$$\nabla \cdot \underset{\sim}{u} = 0 \qquad \text{(b)}$$

$$(2)$$

where

$$C = \begin{bmatrix} 0 & -2r^{-1}V & 0 \\ D_*V & 0 & 0 \\ DW & 0 & 0 \end{bmatrix} \qquad \text{(3a)}$$

$$D = \partial/\partial r, \quad D_* = D + 1/r \qquad \text{(3b)}$$

$$\nabla = D_* \underset{r}{\varepsilon} + r^{-1}\partial/\partial\theta\ \underset{\theta}{\varepsilon} + \partial/\partial z\ \underset{z}{\varepsilon} \qquad \text{(3c)}$$

$$\hat{\nabla} = \nabla - \partial/\partial z\ \underset{z}{\varepsilon} \qquad \text{(3d)}$$

and

$$\underset{\sim}{\Xi} = (\Xi_1, \Xi_2, \Xi_3) \qquad \text{(4a)}$$

$$\Xi_1 = -(\underset{\sim}{u}\cdot\nabla u - r^{-1}v^2) \qquad \text{(4b)}$$

$$\Xi_2 = -(\underset{\sim}{u}\cdot\nabla v + r^{-1}uv) \qquad \text{(4c)}$$

$$\Xi_3 = -\underset{\sim}{u}\cdot\nabla w. \qquad \text{(4d)}$$

In (3), $\underset{r}{\varepsilon}$, $\underset{\theta}{\varepsilon}$, $\underset{z}{\varepsilon}$ are unit vectors in the (r,θ,z) directions.

It is convenient to reduce this set of four equations to a single partial differential equation of higher order for the perturbation radial velocity component u. The resulting equation takes the form

$$Lu = \varepsilon N(u) \qquad (5)$$

where

$$L \equiv TM + \hat{\nabla}^4[G^2 + (2VD_*V)/r] \qquad \text{(a)}$$

$$T \equiv G[\hat{\nabla}^2 D + 2r^{-3}\partial^2/\partial\theta^2] + 2r^{-2}V\hat{\nabla}^2\partial/\partial\theta \qquad \text{(b)}$$

$$M \equiv GD_* - r^{-1}D_*V\ \partial/\partial\theta - DW\ \partial/\partial z \qquad \text{(c)}$$

$$(6)$$

$$G \equiv \partial/\partial t + r^{-1}V\ \partial/\partial\theta + W\ \partial/\partial z \qquad \text{(d)}$$

$$N \equiv \hat{\nabla}^4[G\Xi_1 + 2Vr^{-1}\Xi_2] - T\ \underset{\sim}{\nabla}\cdot\underset{\sim}{\Xi} . \qquad \text{(e)}$$

For our purposes, the nonlinear operator N may be thought of as acting on u alone.

We first consider a single plane wave (or normal mode) solution of (5) when $\varepsilon = 0$,

$$u = u(r,\theta,z,t) = A\hat{u}_0(r)E, \qquad E = \exp[i(kz+m\theta-\omega t)] \tag{7}$$

where A is an amplitude parameter, and m (an integer) is the azimuthal wavenumber. Equation (5) is now

$$L(\partial/\partial t,\partial/\partial z,\partial/\partial\theta,\partial/\partial r)(\hat{u}_0 AE) = AE\hat{L}(-i\omega,ik,im,D)\hat{u}_0 = 0. \tag{8}$$

After a little manipulation, $\hat{L}$ can be written in the form

$$\hat{L}(\ ) = S^{-2}\{\gamma^2 D[SD_*(\ )]-[\gamma^2+\gamma m^{-1}a(r)+b(r)](\ )\} \tag{9}$$

where

$$S \equiv r^2(m^2+k^2r^2)^{-1}, \qquad \gamma \equiv kW + mV/r-\omega$$
$$a \equiv mrD[Sr^{-1}D\gamma + mr^{-3}V], \qquad b \equiv -2kr^{-2}VS[krD_*V-mDW] \tag{10}$$

The operator in the $\{\ \}$ in (9) was derived by Howard and Gupta [14]. The solution of equation (8) provides the dispersion relation $\omega = \omega(k,m)$ and the modal structure $\hat{u}_0(r)$.

Instability of Columnar Vortices

The determination of the stability properties of vortices, even those with the symmetry of columnar vortices, is difficult and few results are available that can be applied to general velocity profiles. If attention is restricted to axisymmetric disturbances only, then Howard and Gupta [14] provide a necessary condition for instability, generalizing a famous result of Rayleigh's. Unfortunately, vortices often are less stable to nonaxisymmetric disturbances, and the Howard-Gupta criterion is therefore not usually the significant one. Ludwieg [15] has given a stability criterion for nonaxisymmetric perturbations of swirling flow in a narrow annular gap. It was hoped that the criterion would indicate local instability for general vortices, but it does not turn out to accomplish this; it often predicts instability of flows that are in fact stable.

Recently, Professor Keith Stewartson and I have developed [16] a criterion for instability of columnar vortices to nonaxisymmetric disturbances that is easily applied to general velocity profiles in unbounded fluids. The argument, which I shall briefly sketch, leads to the following condition sufficient for instability

$$2VD\Omega[D\Omega \cdot D(rV) + (DW)^2] < 0. \tag{11}$$

Apart from a numerical factor, this condition agrees with Ludwieg's [15].
The two criteria were derived, however, in entirely different ways, and the
differing numerical factor is crucial; flows which Ludwieg's criterion
declare to be unstable and are observed experimentally to be stable, are
not predicted to be unstable by criterion (11).

The condition (11) was an outgrowth of the development of a method of
calculating the growth rates in unstable vortices in an unbounded fluid.
From equations (8) and (9), the radial velocity component associated with
infinitesimal perturbations of the vortex must satisfy the Howard–Gupta
equation

$$D\left[SD_{*}u\right] - \left\{1 + a(r)/(m\gamma) + b(r)/\gamma^{2}\right\}u = 0, \tag{12}$$

and homogeneous boundary conditions at the axis $r = 0$ and at infinity. The
coefficients a and b are the complicated functions of the two profiles of the
basic velocity field and their derivatives and of the wavenumbers k and m
previously given. The physical interpretation of a is not clear, but b
represents the rate of stretching of the basic flow in a direction normal to
the wavefronts of the perturbation illustrated in Figure 4.

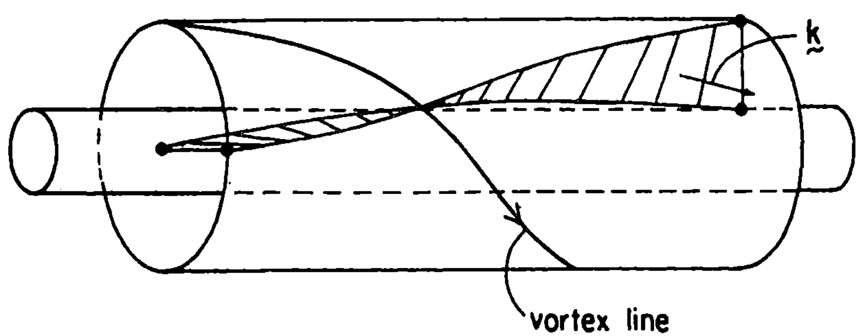

Figure 4.   Advancing wavefront for nonaxisymmetric waves.

Experimental work on vortex breakdown in tubes reported in Faler and
Leibovich [9,10], Garg & Leibovich [17], and Leibovich [1] show that vortices
both upstream and (well) downstream of the reversed flow region are well
fitted by the particular vortex model (Figure 5)

$$V(r) = q\left[1 - \exp(-r^{2})\right]/r, \quad W(r) = \exp(-r^{2}) \tag{13}$$

which is given here in dimensionless form. The stability of this particular
flow has been investigated by Lessen, Singh, and Paillet [18] and by Duck and
Foster [19] by numerical integration of (12). They find that the flow is
more unstable for negative values of m. For fixed (unstable) values of the
swirl parameter q, the maximum growth rates increase as $-m$ increases, at
least for $-m = 1,2,\ldots,6$. Since there is nothing unusual about the
particular form (13), one expects that the stability characteristics it
exhibits will be shared by a wide class of vortex flows. This immediately
suggests that the behavior for large positive values of $n \equiv -m$ needs to be
understood, and this was our aim in [16].

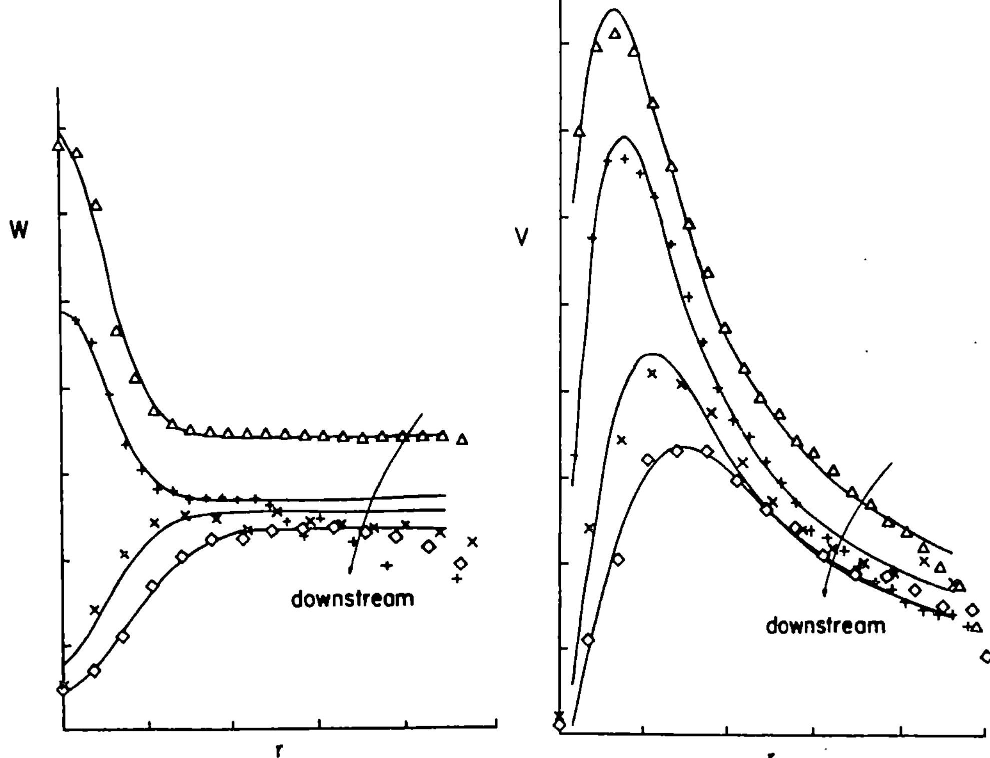

Figure 5. Laser doppler anemometer traverses taken by A.K. Garg at various axial stations in a flow containing a spiral breakdown. The order of progression from upstream to downstream is indicated. Symbols are experimental data, curves are fits of the form (13).

A transformation brings (12) into a standard form

$$D^2\phi = K(r) \cdot \phi \, , \tag{14}$$

and ϕ, which is just proportional to u, also satisfies homogeneous boundary conditions. Our interest now centers on the possibility of unstable modes which might exist in the limit $n \to \infty$ $(m \to -\infty)$. The function K in (14) is given by

$$K = n^2\{1+a/m\gamma+b/\gamma^2+[1+10\beta^2r^2-3\beta^4r^4]/[4n^2(1+\beta^2r^2)^3]\} \equiv n^2 J(r), \tag{15}$$

$$\beta \equiv k/n$$

and for fixed k and n, $K \to \infty$ as $r \to 0$ and $K \to k^2$ as $r \to \infty$, provided the vorticity vanishes as $r \to \infty$. A little exploration shows that in these cases, typified by our prototype (13), there are no solutions as $n \to \infty$ if k is fixed; we therefore treat the problem with β fixed as $n \to \infty$.

Now a general necessary condition for the existence of solutions of
second order differential equations with complex coefficients in the form of
equation (14) and homogeneous boundary conditions is

$$\text{real } (K) < 0 \tag{16}$$

in some interval in the domain of interest. To the best of our knowledge,
this condition has not been previously discovered. When applied to the
Rayleigh equation for plane parallel flow, for example, this condition gives
a strengthened version of the Fjortoft stability criterion: this new version
is

$$k^2 + U''(U-c_r)/\left|U-c\right|^2 < 0. \tag{17}$$

Applied to the vortex problem, condition (16) implies the existence of a
minimum of real(K) at a point r_0, where $0 < r_0 < \infty$, at which real(K) <
0. In the limit $n \to \infty$, this requires J to have a double zero at $r = r_0$,
and this allows the determination of the asymptotic value of the eigenvalue
(ω_r, ω_i), where ω_r and ω_i are the real and imaginary parts of the
complex frequency. One finds the asymptotic growth rate

$$\omega_i = \sqrt{b_o} \tag{18a}$$

and real frequency

$$\omega_r = n\left[\beta W - \Omega\right] , \quad \Omega = V/r \tag{18b}$$

provided that, at wavenumber ratio $\beta = k/n$, there exists a root r_0 of the
equation

$$D(\beta W - \Omega) = 0, \tag{19}$$

at which

$$b_o > 0. \tag{20}$$

This condition has a kinematic interpretation. Condition (19) implies that
the wavefront is parallel to that principal strain axis in the basic vortex
flow corresponding to zero rate of extension. At that level, b_0, which
represents the local rate of stretching of vortex lines in the basic flow by
perturbations normal to the wavefront, must be positive.

Combining (19) and (20), one arrives at condition (11) as a sufficient
condition for instability. If (11) is satisfied, then we can certainly
assert that the flow is unstable since it is unstable to disturbances with
sufficiently large azimuthal wavenumbers. Numerical experience also suggests
that under these circumstances, the flow will be unstable to perturbations at
all positive n. It is worth noting that conditions that are sufficient for
instability of parallel flows are very rare.

If (11) is applied to experimentally realized flows well upstream
vortex breakdown, we find that (11) is _not_ satisfied upstream in any of the

data reviewed in [1], so we cannot assert that the flow is unstable there.
The wakes well downstream of all the bubble breakdowns are unstable by this
criterion, however, and the wakes of all spiral breakdowns are either
unstable, or very close to unstable, according to (11). It is important to
emphasize that (11) is known not to be necessary for instability. Stewartson
and I have computed unstable modes for n = 2,3,4,5 for the flows (13) for
values of q which do not satisfy (11), and Lessen et al [18] allude to
similar calculations. These modes are typically close to marginal, and it
seems likely that under these circumstances, the most unstable disturbances
correspond to n = 1. It appears as if all data available shows the wakes of
all vortex breakdowns are unstable to non-axisymmetric disturbances, but that
they are stable to axisymmetric ones. Since the real frequency turns out to
to be negative for flows in the form (13), which fit the experimental data
well, the theory outlined above predicts prograde unstable modes.

Finite Amplitude Wave Propagation: Axisymmetric Waves

We now consider stable systems. These allow waves to propagate, and
infinitesimal wavy disturbances in the form (7) are possible. The
description can be extended to account for finite amplitude effects by taking

$$u(r,\theta,z,t) = u_o + \varepsilon u_1 + \varepsilon^2 u_2 + \ldots$$

and to allow for modulation of the wave on a long spatial scale $(\varepsilon^{-1}R)$ by
a multiple scale argument. This will force a slow modulation in time as
well. Thus we assume

$$u_o = u_o(r,\theta,z,X,t,\tau_1,\tau_2)$$

where

$$X = \varepsilon z$$

and slow time modulation is accounted for by

$$\tau_1 = \varepsilon t.$$

We have included an additional slow time variable

$$\tau_2 = \varepsilon^2 t$$

required to balance higher order modulation and nonlinear effects. The slow
time and space variations are accommodated by permitting the amplitude A to
depend upon X,τ_1, and τ_2, thus

$$u_o = \hat{u}_o(r)A(X,\tau_1,\tau_2)E + c.c. \tag{21}$$

where $\hat{u}_o(r)$ satisfies the Howard-Gupta equation (12).

If axisymmetric modes (m=0) are considered, it is found (Benjamin [20],
Leibovich [21]) that the axial phase speeds are greatest for waves of extreme
length (k→0). In this case, ω→0 as well and $c_o = \omega/k$ is finite. The
amplitude modulation in this case is governed by the Korteweg-deVries
equation

$$\partial A/\partial \tau_1 + (c_o + \varepsilon c_1 A)\partial A/\partial X + \varepsilon c_2 \partial^3 A/\partial X^3 = 0, \qquad (22)$$

if the flow is confined to a tube; the constant coefficients c_1 and c_2 are computed as othogonality conditions. In the case of a radially unbounded flow, the third derivative is replaced by an integral over A [22]: the behavior of solutions of the integro-differential equation are similar [23] to those of (22).

The solutions of primary interest are the waves of permanent form; for (24), this is the soliton

$$A = \alpha \, \mathrm{sech}^2\left[(c_1\alpha/3c_2)^{1/2}(X-(c_o-\alpha\varepsilon c_1)\tau_1)/2\right]$$

If ε is regarded now as arbitrarily adjustable parameter (instead of a small one), then one finds [22] that if it is large enough recirculation regions form with a shape (bubble fineness ratio) similar to those seen experimentally.

If the motion is slightly dissipative and takes place in a vortex with a weak axial variation, equation (24) is modified and becomes

$$\partial A/\partial \tau_1 + (c_o + \varepsilon c_1 A)\partial A/\partial X + \varepsilon c_2 \partial^3 A/\partial X^3 = \varepsilon c_3 A \qquad (23)$$

where c_0, c_1, c_2, c_3 all depend weakly upon axial distance [13]. If the wave is required to be stationary in a laboratory frame of reference, one finds that ε is no longer a free parameter, but is restricted to a narrow range of values, none of which can be small! In this range, the wave is trapped in a definite interval along the vortex axis and contains a region of reversed flow. The model therefore implies that the phenomenon is necessarily of large amplitude and that it is localized, with position determined by the basic vortex. Of course, the theory is not self-consistent when ε is not small – we have forced the weakly nonlinear theory beyond its limits of logical validity – but the resulting model conveys information that a self-consistent weakly nonlinear theory cannot.

Finite Amplitude Waves: Nonaxisymmetric Disturbances

The spiral form of vortex breakdown cannot be addressed by an axisymmetric theory. Hasimoto [24] has explored the possibility of a large amplitude twisting of an otherwise straight and infinitely long vortex filament. We believe that his theoretical work may have application to spiral vortex breakdowns, and we currently are investigating the possibility.

Hasimoto uses the local induction forms of the Biot-Savart law to compute motion of the filament, which he finds to be governed by the nonlinear Schrodinger equation (NSE). This approximation neglects the contribution to the motion of a point on the filament made by parts of the vortex at distances larger than those for which the curvature of the filament can be regarded as sensibly constant, and it also ignores length scales of the order of the core radius or smaller. The result of this is a theory that cannot be directly related to the properties of the vortex.

An alternate route is weakly nonlinear theory, which also yields a nonlinear Schrodinger equation. It is my feeling that the weakly nonlinear theory can be used to "calibrate" the Hasimoto model, that is, to supply the missing links with the vortex structure. Then the Hasimoto model can be used with some justification to compute spiral deformations of the filament that are not of small amplitude. I outline here a schematic of the derivation of the NSE according to weakly nonlinear theory.

Return to (21) and allow m to be an integer, and do not restrict considerations small k. Modulation effects are most easily accounted for following the method of Newell [24]. With $\partial/\partial t$ and $\partial/\partial z$ replaced in L by

$$\partial/\partial t + \varepsilon\partial/\partial\tau_1 + \varepsilon^2\partial/\partial\tau_2, \quad \partial/\partial z + \varepsilon\partial/\partial X \tag{24}$$

we expand the linear operator in equation (5). The calculation is facilitated by utilizing the correspondence between the operator L and its Fourier transform with respect to the variables t and z, expressed in (9), which permits the identification

$$ik = ik_o + \varepsilon\partial/\partial X, \quad -i\omega = -i\omega_o + \varepsilon(\partial/\partial\tau_1 + \varepsilon\partial/\partial\tau_2 + \ldots),$$

when $ik_o{\sim}\partial/\partial z$, the "fast" spatial scale, and $-i\omega_o{\sim}\partial/\partial t$, the "fast" time scale. Omitted terms indicate additional slow times required to extend the analysis to higher order. Thus

$$
\begin{aligned}
L(\partial/\partial t &+ \varepsilon\partial/\partial\tau_1 + \varepsilon^2\partial/\partial\tau_2 + \ldots, \partial/\partial z + \varepsilon\partial/\partial X, \partial/\partial\theta, D) \\
&= L_o + L_\omega(\omega-\omega_o) + L_k(k-k_o) + \frac{1}{2}L_{\omega\omega}(\omega-\omega_o)^2 \\
&\quad + L_{\omega k}(k-k_o)(\omega-\omega_o) + \frac{1}{2}L_{kk}(k-k_o)^2 + \ldots
\end{aligned}
\tag{25}
$$

here L_o is L based upon the fast variables only and L_ω represents the derivative of the double Fourier transform of L with respect to ω, the result then inverted to produce operators with respect to the fast variables only. Higher derivatives have the same interpretation, consequently

$$
\begin{aligned}
L(\partial/\partial t &+ \varepsilon\partial/\partial\tau_1 + \varepsilon^2\partial/\tau_2 + \ldots, \partial/\partial z + \varepsilon\partial/\partial X, \partial/\partial\theta, D) = \\
L(\partial/\partial t &, \partial/\partial z, \partial/\partial\theta, D) + i\varepsilon\left[L_\omega\partial/\partial\tau_1 - L_k\partial/\partial X\right] + \varepsilon^2\left[iL_w\partial/\partial\tau_2 - \frac{1}{2}L_{kk}\partial^2/\partial X^2 \right. \\
&\left. + L_{\omega k}\partial^2/\partial\tau_1\partial X - \frac{1}{2}L_{\omega\omega}\partial^2/\partial\tau_2^2\right] + \ldots
\end{aligned}
\tag{26}
$$

We now introduce an expansion of u in powers of ε

$$u = u_o + \varepsilon u_1 + \varepsilon^2 u_2 + \ldots \tag{27}$$

The nonlinear operator N is quadratic in u, and we can also write $N(u_o + \varepsilon u_1 + \varepsilon^2 u_2 + \cdots)$ as a series in ε,

$$N(u) = N_o(u_o) + \varepsilon N_1(u_o, u_1) + \ldots \tag{28}$$

Thus, (5) reduces to a sequence of equations, the first three of which are

$$L_o u_o = 0 \qquad\qquad (a)$$

$$L_o u_1 = N_o(u_o) - i\left[L_\omega \partial u_o/\partial\tau_1 - L_k \partial u_o/\partial X\right] \qquad\qquad (b)$$

$$L_o u_2 = N_1(u_o,u_1) - i\left[L_\omega \partial u_1/\partial\tau_1 - L_k \partial u_1/\partial X\right] \qquad\qquad (29)$$
$$- \left[iL_{\omega_2} \partial u_o/\partial\tau_2 - \frac{1}{2} L_{kk} \partial^2 u_o/\partial X^2 \qquad\qquad (c)\right.$$
$$\left. + L_{\omega k} \partial^2 u_o/\partial\tau_1 \partial X - \frac{1}{2} L_{\omega\omega} \partial^2 u_o/\partial\tau_1^2\right] \ .$$

From (7) and (29b), one sees that u_1 satisfies the equation

$$L_o u_1 = N_o - iE\left[\partial A/\partial\tau_1 \hat{L}_\omega \hat{u}_o - \partial A/\partial X \hat{L}_k \hat{u}_o\right] \qquad\qquad (30)$$

where

$$N_o = iP_o(r)A^2E^2 + \text{c.c.} \qquad\qquad (31)$$

with P_o real if $\hat{u}_o$ is real. Particular solutions for u_1 are of the form

$$u_1 = iu_1^{(1)}EA_X + iu_1^{(2)}A^2E^2 + \text{c.c.} \qquad\qquad (32)$$

For the expansions to be well-ordered for large t, we must suppress secular terms arising from the bracket in (20). This requires the orthogonality condition

$$\partial A/\partial\tau_1 \int_{R_1}^{R_2} r\hat{u}_o^+ \hat{L}_\omega \hat{u}_o \, dr - \partial A/\partial X \int_{R_1}^{R_2} r\hat{u}_o^+ \hat{L}_k \hat{u}_o \, dr = 0 \ . \qquad\qquad (33)$$

Here $\hat{u}_o^+$ is the solution to the equation adjoint to $\hat{L}_o\hat{u} = 0$, and is given by

$$\hat{u}_o^+ = \gamma^{-2}s^2\hat{u}_o \ . \qquad\qquad (34)$$

Equation (33) defines the group velocity c_g of the wave packet entered at wavenumber k_o,

$$c_g = -\langle\hat{u}_o^+\hat{L}_k\hat{u}_o\rangle/\langle\hat{u}_o^+\hat{L}_\omega\hat{u}_o\rangle \qquad\qquad (35)$$

where we have used the notation

$$\langle f\rangle = \int_{R_1}^{R_2} rf dr. \qquad\qquad (36)$$

The expression (35) can be shown to be

$$c_g = -\langle Q_1 \hat{u}_o^2 \rangle / \langle Q_2 \hat{u}_o^2 \rangle \qquad (37)$$

where Q_1 and Q_2 are real for ω and k real.

Adopting henceforth a coordinate system moving with the packet

$$z = X - c_g \tau_1 \qquad (38)$$

we find that $u_1^{(1)}$ and $u_1^{(2)}$ are real for ω real, and satisfy the inhomogeneous equations

$$\hat{L}_o \hat{u}_1^{(1)} = c_g \hat{L}_\omega \hat{u}_o + \hat{L}_k \hat{u}_o \qquad (a)$$

$$\hat{L}_1 \hat{u}_1^{(2)} = P_o(r) \qquad (b) \qquad \qquad (39)$$

where

$$\hat{L}_1 \equiv L(-2i\omega, 2ik, 2im, D). \qquad (40)$$

Proceeding to the next order, we find

$$N_1 = P_1(r)A|A|^2 E + P_2(r)AA_z E^2 + P_3 A|A|^2 E^3 + c.c. \qquad (41)$$

where the P_i are real for ω real. Consequently (29c) dictates a particular solution u_2 of the form

$$u_2 = \hat{u}_2^{(1)}(r)A|A|^2 E + \hat{u}_2^{(2)} AA_z E^2 + u_2^{(3)} A|A|^2 E^3 + c.c. \qquad (42)$$

and the orthogonality condition

$$i\langle \hat{u}_o^+ \hat{L}_\omega \hat{u}_o \rangle \partial A/\partial \tau_2 - \frac{1}{2}\langle \hat{u}_o^+(\hat{L}_{kk}\hat{u}_o + 2c_g \hat{L}_{\omega k}\hat{u}_o + c_g^2 \hat{L}_{\omega\omega}\hat{u}_o)\rangle \partial^2 A/\partial z^2 - \langle \hat{u}_o^+ P_1 \rangle A|A|^2 = 0 \qquad (43)$$

is required for $\hat{u}_2^{(1)}$ to exist. If none of the coefficients vanish, then (43) is the cubic nonlinear Schrödinger equation, which we write in the form

$$i\partial A/\partial \tau_2 + \mu \partial^2 A/\partial z^2 + \nu A|A|^2 = 0 . \qquad (44)$$

If $\mu\nu > 0$, soliton solutions of the NSE exist, and a single packet of vortex twisting can propagate without change of form. For representative vortex core structures, Huiyang Y. Ma has calculated μ and ν, and finds that this condition is satisfied.

The next step is to use this weakly nonlinear theory to calibrate Hasimoto's solutions, and Ma and I are working on this question. In this connection, only $|m| = 1$ is of interest, since only this mode corresponds to a bending of the filament off of the axis.

Nonlinear interactions between axisymmetric and nonaxisymmetric modes are of considerable interest. The bubble form of vortex breakdown is definitely not axisymmetric, and it may well be that the spiral form has an axisymmetric component. Thus the interaction of axisymmetric and nonaxisymmetric waves is expected to be important. A beginning has been made by Huang [26], but this work has yet to be reduced to a form suited to a theoretical model of physical phenomena.

A Large Amplitude Long Wave Model

Existing linear or weakly nonlinear wave propagation theoreies are useful models that seem to describe a number of features that are observed in vortex breakdown flows. The successes of these theories when applied to the strongly nonlinear processes occurring in vortex breakdown flows suggests that there may be wave propagation possibilities of arbitrary amplitude. I will present here an analysis which, although presently in an incomplete form, indicates the possibility of finding an axisymmetric wave propagation phenomenon of arbitrary amplitude.

A streamfunction is available for axisymmetric motions, it is convenient to adopt the formulation used in [22]. If $\kappa = R/L$ is the ratio of typical radial length scales of variation to typical axial scales, then the exact (inviscid) equations governing axisymmetric motions may be brought into the following dimensionless form (see [22] for scalings used),

$$\zeta_t + 2\psi_y \zeta_z + y^{-1}(\Gamma^2)_z - 2y\psi_z [y^{-1}\zeta]_y = 0. \qquad (a)$$

$$\Gamma_t - 2\psi_z \Gamma_y + 2\psi_y \Gamma_z = 0. \qquad (b)$$

$$\zeta \equiv 4y\psi_{yy} + \kappa^2 \psi_{zz} \qquad (c) \qquad\qquad (45)$$

$$\Gamma \equiv rv \qquad (d)$$

$$y \equiv r^2. \qquad (e)$$

Thus Γ is normalized circulation about the symmetry axis z, and ζ is the (negative of) the azimuthal vorticity; ψ is the usual Stokes streamfunction, but expressed in terms of y rather than r.

We now suppose that $\kappa^2 \ll 1$ and neglect terms proportional to it. Let

$$\psi = \phi(y)A(z,t) \qquad (46)$$

$$\Gamma = 2g(y)\phi A$$

and substitute into (45). Next, take

$$A_t = 2\lambda AA_z \quad , \qquad (47)$$

where λ is a constant and the form of A is given as a function of z at $t = 0$. The coefficients ϕ and g satisfy

$$\phi\phi''' - \phi''(\phi'+\lambda) - \phi^2 g^2/y^2 = 0.$$

$$\phi g' - \lambda g = 0 \qquad\qquad (48)$$

$$(\ )' \equiv d/dy.$$

subject to boundary conditions at $y = (y_i, y_2)$ (where $y_1 = 0$ if there is no inner boundary)

$$\phi(y_1) = \phi(y_2) = 0 \qquad\qquad (49)$$

and a normalization condition

$$\phi'(y_1) = \alpha, \ g(y_1) = \beta \qquad\qquad (50)$$

The mathematical problem outlined here, if solutions exist, describes the propagation of fully nonlinear long waves. If $A \rightarrow$ constant as $|z| \rightarrow \infty$, the problem describes the distortion of a columnar vortex whose structure is given by ϕ and g. Since these functions cannot be arbitrarily prescribed, the model defines a special class of vortices on which arbitrarily large deformations of this form are admissible. Given the boundary conditions (49), the differential equations (48) are singular at the end points y_1 and y_2, and viscous effects may prove to be important near these boundaries. When $\lambda A_z(z,0) > 0$ in some z-interval, the solution will break down in finite time, a process reminiscent of shock formation following the nonlinear steepening of simple waves in gasdynamics, and suggestive of a vortex breakdown event. Wave breaking here requires that dispersive effects associated with the finite length of a real wave be restored in order to establish the structure near the wavefront.

Since the form of the solution outlined here requires ϕ to vanish at the boundaries y_1 and y_2, it represents a flow with zero volume rate of flow. This can be adjusted at will, of course, by making the Galilean transformation

$$z_1 = z + \upsilon t, \ \Psi(y, z_1, t) = \upsilon y + \psi$$

thereby producing a flow with an arbitrarily adjustable mass flow rate.

Conclusion

Vortex breakdowns take place in many flows of practical importance, and this naturally makes it worth our while to try to understand why it happens, and what to expect when it occurs. The problem also has theoretical interest independent of its practical importance. It is a paradigm of events that occur in a number of other flows which are physically quite unrelated. The occurrence of propagating regions containing velocity reversals, for example, is a known feature in density stratified flows. Another example of parallel phenomenon is the turbulent spot in wall-bounded shear flows and other "coherent structures" in turbulent flows. Vortex breakdown is generally followed by a transition to turbulence; because it is so well localized, it can be regarded as a switch that turns on (or turns up) the turbulence.

I have outlined here three of the theoretical research directions
currently being pursued in efforts to better understand the highly nonlinear
processes leading to vortex breakdown. Two of these approaches are linear or
weakly nonlinear, with the present focus on incorporating nonaxisymmetric
disturbances. Nonaxisymmetric motions are known to occur in all vortex
breakdowns (at least as the term has been defined here), and to dominate the
"spiral" form. The third line of investigation discussed here is the search
for a fully nonlinear theory. A start on a theory, restricted to
axisymmetric motions, has been made. This takes a form analogous to
nonlinear shallow water theory or nonlinear one dimensional unsteady
gasdynamics, but is rather more complicated than either of these classical
theories.

References

1. Leibovich, S., 1978. The structure of vortex breakdown, Ann. Rev. Fluid
    Mechanics 10, 221-46.

2. Ludwieg, H., 1970. Vortex breakdown, Dtsch. Luft Raumfahrt Rep. 70-40.

3. Hall, M.G., 1972. Vortex breakdown, Ann. Rev. Fluid Mech. 4, 195-218.

4. Kopecky, R.M. and K.E. Torrance, 1973. Initiation and structure of
    axisymmetric eddies in a rotating stream. Comput. Fluids 1, 289-300.

5. Grabowski, W.J. and S.A. Berger, 1976. Solutions of the Navier-Stokes
    equations for vortex breakdown. J. Fluid Mech. 75, 525-44.

6. Vogel, H.U., 1968. Max-Planck Institut fur Stromungsforschung,
    Gottingen, Bericht 6.

7. Ronnenberg, B., 1977. Max-Planck-Institut fur Stromungsforschung,
    Gottingen, Bericht 20.

8. Lugt, H.J. and Haussling, H.J., 1982. Axisymmetric vortex breakdown in
    rotating fluid within a container, to appear in J. Applied Mech.

9. Faler, J.H. and S. Leibovich, 1977a. Disrupted states of vortex flow
    and vortex breakdown, Phys. Fluids 20, 1385-1400.

10. Faler, J.H. and S. Leibovich, 1977b. An experimental map of the
    internal structure of a vortex breakdown, J. Fluid Mech. 86, 312-335.

11. Sarpkaya, T. 1971a. On stationary and travelling vortex breakdowns, J.
    Fluid Mech. 45, 545-59.

12. Escudier, M.P. and Zehnder, N. 1982. Vortex flow regimes, J. Fluid
    Mech. 115, 105-121.

13. Randall, J.D. and S. Leibovich, 1973. The critical state: a trapped
    wave model of vortex breakdown, J. Fluid Mech. 53, 495-515.

14. Howard, L.N. and A.S. Gupta, 1962. On the hydrodynamic and
    hydromagnetic stability of swirling flows, J. Fluid Mech. 14, 463-76.

15. Ludwieg, H., 1961. Ergäzung zu der Arbeit: "Stabilität der Stromung in
    einem zylindrischen Ringraum", Z. Flugwiss 9, 359-361.

16. Leibovich, S. and K. Stewartson, 1982. A sufficient condition for the instability of inviscid columnar vortices, to be published in J. Fluid Mech.

17. Garg, A.K. and S. Leibovich, 1979. Spectral characteristics of vortex breakdown flowfields, Phys. Fluids, 22, 2053-64.

18. Lessen, J., P.J. Singh and F. Paillet, 1974. The stability of a trailing line vortex, J. Fluid Mech. 63, 753-63.

19. Duck, P.W. and M.R. Foster, 1980. The inviscid stability of a trailing line vortex, Z.A.M.P., 31, 523-30.

20. Benjamin, T.B., 1962. Theory of the vortex breakdown phenomenon, J. Fluid Mech. 14, 593-629.

21. Leibovich, S., 1979. Waves in parallel or swirling stratified shear flows. J. Fluid Mech. 93, 401-412.

22. Leibovich, S., 1970. Weakly nonlinear waves in rotating fluids. J. Fluid Mech. 42, 803-822.

23. Leibovich, S. and Randall, J.D., 1972. Solitary waves in concentrated vortices, J. Fluid Mech. 51, 625-635.

24. Hasimoto, H., 1972. A soliton on a vortex filament, J. Fluid Mech. 51, 477-485.

25. Newell, A.C., 1972. The post bifurcation stage of baroclinic instability, J. Atm. Sci., 29, 64-76.

26. Huang, J-H., Nonlinear interaction between spiral and axisymmetrical disturbances in vortex breakdown, Ph.D. Thesis, Cornell University, 1974, 140 pp.

WIDERSTANDSREDUZIERUNG BEI KRAFTFAHRZEUG-
ÄHNLICHEN KÖRPERN

von

H. Ludwieg

Deutsche Forschungs- und Versuchsanstalt
für Luft- und Raumfahrt e.V.
Aerodynamische Versuchsanstalt Göttingen
Institut für Experimentelle Strömungsmechanik

1. Einleitung

Betrachtet man den Widerstandsbeiwert (C_W-Wert) von Kraftfahrzeugkarosserien, so liegt dieser bei der heute üblichen Pontonform bei etwa $C_W \doteq 0,4$. Der Einfluß der recht unterschiedlichen Heckformen wie Stufenheck, Fließheck, Kastenheck und abgeschrägtes Kastenheck auf den C_W-Wert ist dabei nur gering, wie Messungen des C_W-Wertes von R. Buchheim, K.R. Deutenbach und H.J. Lückoff zeigen [1].

Beim Widerstand eines Fahrzeuges kann man nun zwei Anteile unterscheiden. Einmal den Widerstand einer idealisierten Grundform, im folgenden kurz als Grundform bezeichnet, und zum anderen den zusätzlichen Widerstand, der durch die verschiedenen Anbauten und Einsprünge an der Karosserie entsteht, wie z.B. Räder, Radkästen, Kühlereinlauf, Stoßstangen, Auspuffanlage, Außenspiegel, einspringende Fenster, Dachrinnen usw.. Der zweite Anteil läßt sich nur durch konstruktive Einzelmaßnahmen senken und soll hier nicht weiter behandelt werden. Beide Widerstandsanteile sind bei den heute üblichen Karosserien etwa gleich groß, d.h. die Grundform besitzt einen C_W-Wert von etwa 0,2. Nur dieser Anteil soll im folgenden näher betrachtet werden.

2. Allgemeine Betrachtungen zum Fahrzeugwiderstand

Betrachtet man nun den Widerstand der Grundform mit einem C_W-Wert von etwa 0,2, so läßt sich dieser Widerstand in einen Reibungswiderstand und einen Druckwiderstand zerlegen. Der Anteil des Reibungswiderstandes ist dabei verhältnismäßig klein, etwa 15-20 % vom Gesamtwiderstand, und läßt sich durch die Formgebung der Karosserie nur unwesentlich beeinflussen. Eine wesentliche Senkung des Widerstandes läßt sich also nur durch Senkung des Druckwiderstandes erreichen.

Dieser Druckwiderstand entsteht nun einmal als sogenannter Totwasserwiderstand durch das Abreißen der Strömung an der senkrecht zur Fahrtrichtung stehenden Heckfläche und auch an der nur mäßig geneigten Heckscheibe bei Fahrzeugen mit Stufenheck oder abgeschrägtem Kastenheck. Auch bei Fließheckfahrzeugen mit relativ

steil gestellter Heckscheibe tritt diese Form der Ablösung auf.
Zum anderen entsteht auch durch die bekannten, dreidimensionalen
Wirbelablösungen an der oberen Fahrzeuglangskante im Bereich der
Frontscheibe (A-Säulenablösung) und im Bereich der Heckscheibe
von Fließheckfahrzeugen (C-Säulenablösung) ein Druckwiderstand,
der hier kurz als Wirbelwiderstand bezeichnet sei. Diese Art der
Ablösung bedingt die häufig beobachteten und untersuchten starken
Längswirbel im Nachlauf der Fahrzeuge.

Es sei an dieser Stelle aber auch gleich darauf hingewiesen, daß auch bei stumpfen Heckformen mit voller Heckablösung,
wie Stufenheck- oder gar Kastenheckfahrzeugen, Längswirbel mit
beträchtlicher Intensität im Nachlauf auftreten können [2]. Zusätzlich tragen noch die Ablösung im Stau der Frontscheibe und
eventuelle Ablösungen an zu wenig abgerundeten Querkanten des
Fahrzeuges zum Druckwiderstand bei.

Versucht man nun eine quantitative Aufteilung des Druckwiderstandes in die beiden Hauptanteile Totwasser- und Wirbelwiderstand bei einer vorgegebenen Fahrzeugform, so stößt man auf
erhebliche Schwierigkeiten. Selbst eine genaue Messung der Druckverteilung auf der gesamten Fahrzeugoberfläche im Windkanal würde
nicht viel nützen, da einerseits die Totwasserbildung auch die
Druckverteilung auf den Flächen vor der Ablösung beeinträchtigt
und da andererseits auch die abgehenden Wirbel das Totwasser beeinflussen.

Ein Versuch eine solche Aufteilung durchzuführen, wurde
von S.R. Ahmed [2] durchgeführt. Hierbei wurde im Windkanal in
verschiedenen Ebenen hinter dem Fahrzeug eine Dreikomponenten-
geschwindigkeitsmessung im Nachlauf vorgenommen. Ein "Quasi-
Wirbelwiderstand" wurde dann dadurch definiert, daß man den Fluß
an kinetischer Energie der Querkomponente der Strömungsgeschwin-
digkeit durch eine Kontrollfläche als Schleppleistung bezüglich
des Wirbelwiderstandes ansieht. Dies kann aber auch nur einen
rohen Hinweis auf die Größe des Wirbelwiderstandes liefern, denn
der Fluß an kinetischer Energie der Querströmung ändert sich mit
dem Abstand der Kontrollebene vom Fahrzeug durch Wirbeldiffusion
und durch Änderung des statischen Druckes. Die Messungen zeigen
daher auch eine starke Änderung des "quasi-Wirbelwiderstands"
mit dem Abstand der Meßebene vom Modell, so daß keine eindeutige
Aussage über die Größe des "Quasi-Wirbelwiderstandes" möglich ist.

Auch der mehrfach vorgenommene Versuch, den Druckwiderstand in Analogie zur Tragflügeltheorie, in einen auftriebsun-
abhängigen und in einen auftriebsabhängigen Anteil aufzuteilen,
wobei der zweite Anteil dann ja ein Wirbelwiderstand wäre, kann
kaum zu allgemein gültigen Aussagen führen [3]. Man erkennt das
am einfachsten aus der Tatsache, daß ein rotationssymmetrischer,
tropfenförmiger Halbkörper mit Bodenabstand Null in einer Poten-
tialströmung einen erheblichen Auftrieb und trotzdem keinen
Widerstand besitzt.

3. Grundkonzept zur Widerstandsreduzierung

Wir wollen hier gar nicht versuchen, den Druckwiderstand
der Grundform in mehrere Anteile aufzuteilen, was bisher offen-

sichtlich schon an der ungenauen Definition der einzelnen Anteile
scheiterte. Wir wollen aber deutlich zwischen den beiden Haupt-
ursachen für die Widerstandsentstehung, der Totwasserbildung an
den senkrecht oder relativ steilgestellten Heckflächen (Heckablösung) und der Ablösung mit Wirbelbildung bei der seitlichen
Umströmung der Fahrzeuglängskanten (Längskantenablösung) unterscheiden.

Im Prinzip lassen sich nun beide Ursachen für die Widerstandsentstehung durchaus beseitigen; man denke nur an den bereits erwähnten, rotationssymmetrischen, stromlinienförmigen
Halbkörper mit Bodenabstand Null. Er besitzt einen entsprechend
kleinen Widerstand, der im wesentlichen nur aus dem unvermeid-
baren Reibungswiderstand besteht. Allerdings ist diese Form als
Grundform für ein Kraftfahrzeug wenig geeignet.

Wir wollen im folgenden nun untersuchen, ob man allein
durch Vermeidung oder Kleinhaltung der zweiten Ursache (Längskantenablösung) und Beibehaltung realistischerer Grundformen den
Widerstand wesentlich senken kann. Nun benutzt man schon verschiedene Maßnahmen zur Widerstandsreduzierung, die auf dieses
Ziel hinauslaufen. Man rundet die obere Fahrzeuglängskante besonders im Bereich der Front- und im Bereich der Heckscheibe,
stark ab und wählt eine sehr schräg gestellte Frontscheibe und
ein sehr flaches Fließheck. Besonders stark hat man diese Maßnahmen bei aerodynamischen Versuchsfahrzeugen angewandt. Will man
auf diese Weise wesentliche Widerstandsreduzierung erreichen, so
kommt man jedoch auf Formen, die für den praktischen Gebrauch
wenig geeignet sind.

Hier soll nun versucht werden, durch geeignete Wahl des
Breitenverlaufs des Fahrzeuges, die oben geschilderten räumlichen
Ablösungen mit Wirbelbildung zu vermeiden oder zumindest klein
zu halten und dadurch den Widerstand zu senken. Grundsätzlich
kann man den Breitenverlauf so wählen, daß die obere und die unte-
re Fahrzeuglängskante zu Staulinien bzw. Ablöselinien werden, so
daß überhaupt keine seitliche Umströmung dieser Kanten stattfindet und damit natürlich auch keine räumliche Ablösung an den
Kanten auftritt. Dies führt allerdings, wie wir später sehen werden, zu wenig brauchbaren Formen. Andererseits kann man je nach
Stärke der Kantenabrundung eine mehr oder weniger starke Kantenumströmung zulassen, ohne daß eine Ablösung stattfindet. Unter
Ausnützung dieser Tatsache lassen sich dann wesentlich besser
geeignete Formen finden.

4.   Auslegeverfahren der Modelle

Zur Prüfung des in Abschnitt 3 dargelegten Konzeptes sollen
einige Windkanalversuche durchgeführt werden. Um die Modelle hierfür auslegen zu können, müssen wir uns zunächst ein Verfahren
überlegen, das es uns wenigstens näherungsweise gestattet, die
Intensität der Umströmung der Längskanten zu berechnen. Man könn-
te hierfür zwar die sogenannten Pannelverfahren zur Berechnung
des potentialtheoretischen Strömungsfeldes um beliebige Körper
heranziehen, insbesondere das von S.R. Ahmed [3] für Kraftfahrzeugkarosserien entwickelte. Diese Verfahren sind jedoch sehr auf-

wendig und schlecht für die Auslegung von Modellen geeignet, da
sie die Aufgabe behandeln für eine vorgegebene Form die Geschwin-
digkeitsverteilung zu bestimmen.

Zunächst sei die Entstehung der Längswirbel durch räumliche Ablösung an der oberen Fahrzeugkante kurz erläutert. Bei
den heute üblichen Fahrzeugen sind die Querschnitte (Schnitte
senkrecht zur Fahrzeuglängsachse) annähernd Rechtecke oder genauer
gesagt Rechtecke mit leicht nach außen gewölbten Seiten und abgerundeten Ecken. Im Bereich der Frontscheibe, der Heckscheibe
und im schwächeren Maße auch im Bereich des Kühlers und des Koffer-
raumes ändert sich die Höhe dieser Querschnittsflächen verhältnismäßig schnell in Fahrzeugrichtung, während die Breite dieser
Flächen annähernd konstant bleibt. Bei der Umströmung der Karosserie tritt dadurch eine Verdrängungsströmung auf, von der in
Bild 1 die sogenannte Querströmung gezeigt ist (Querströmung ist
die Strömung in einer Querschnittsebene, die entsteht, wenn man
jedem Punkt der Quer-
schnittsebene die in
diese Ebene fallende
Komponente des Strömungs-
vektors zuordnet). Bei
schneller Änderung der
Höhe der Querschnitts-
fläche und nicht sehr
stark abgerundeten oder
gar scharfkantigen Ecken
tritt dabei eine räumli-
che Ablösung mit einer
Trennfläche auf, die sich
zu einem Wirbel aufrollt.
Bild 1 zeigt dabei die
Verhältnisse, wie sie im
Bereich der Frontscheibe
auftreten (Höhenzunahme).
Im Bereich des Hecks
(Höhenabnahme) tritt eine
Umströmung der Kante in
umgekehrter Richtung mit

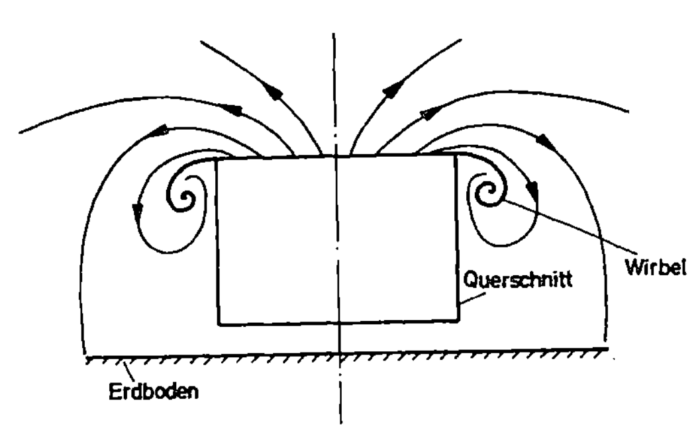

Bild 1: Querströmung in einem Quer-
        schnitt im Bereich der
        Frontscheibe

entsprechender Wirbelbildung auf.

Wie im vorigen Abschnitt bereits erläutert, soll diese
Wirbelbildung dadurch vermieden werden, daß man die Karosserie
so formt, daß der Höhenänderung der Querschnittsform eine ent-
sprechende Breitenänderung zugeordnet wird. Um diese Auslegung
vornehmen zu können, benötigen wir ein Verfahren, das es uns gestattet, bei vorgegebener Höhenänderung des Querschnittes die er-
forderliche Breitenänderung zu berechnen. Wegen der Komplizier-
heit des Strömungsvorganges ist eine genaue Berechnung des Strömungsfeldes unter Berücksichtigung der Grenzschichten, der beschriebenen Längswirbel und des Totwassers nicht möglich.

Für unsere Näherungsrechnungen benutzen wir nun zunächst
das Konzept der sogenannten "slender body"-Theorie, die besagt,
daß bei nicht zu schnellen Querschnittsänderungen in Längsrichtung
die eben erwähnte Querströmung nur vom jeweiligen Querschnitt und
seiner Änderungsgeschwindigkeit in Längsrichtung abhängt und nicht

von den sonstigen Abmessungen der Karosserie. Ferner idealisieren wir die Fahrzeugform so, daß alle Querschnitte exakte Rechtecke sind. Wenn wir dann noch von der Reibung absehen, erhalten
wir für die Querströmung das anhand von <u>Bild 2</u> geschilderte
potentialtheoretische, ebene Strömungsproblem.

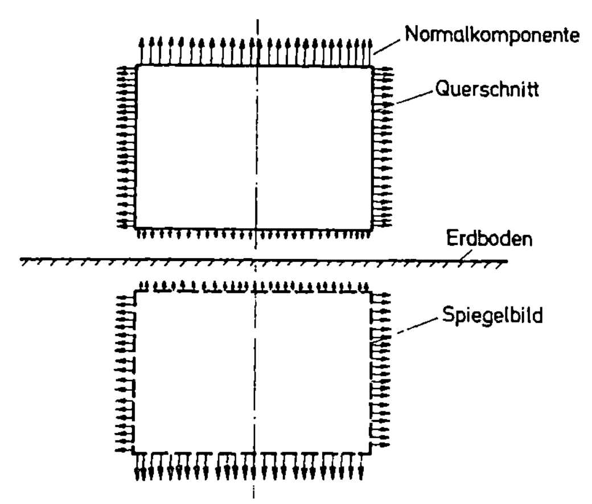

<u>Bild 2</u>: Idealisierte Normalkomponenten-
verteilung in einer Quer-
schnittsebene mit Spiegelbild

Aus dem Rand eines Rechteckes (Querschnitt der Karosserie) tritt auf allen 4 Seiten mit jeweils konstanter Normalkomponente eine Strömung aus, wobei der Betrag der jeweiligen Normalkomponente proportional der entsprechenden Änderung der Querschnittsform in Fahrzeuglängsrichtung ist. Bei ebenem Fahrzeugboden oder bei in Längsrichtung sich nicht ändernder Breite wäre also die Normalkomponente an der Unterseite bzw. an den Seitenflächen gleich Null. Zusätzlich muß die Normalkomponente auf den auf Bild 2 angedeuteten Erdboden verschwinden. Letztere Bedingung läßt sich dadurch ersetzen,
daß man eine am Erdboden gespiegelte Fläche mit gleichen Strö-
mungsbedingungen anordnet. Man erkennt nun sofort, daß bei will-
kürlich gewählter Intensität der konstanten Normalkomponente auf
den vier Seiten eine Umströmung der Ecken der Querschnittsfläche
auftritt, wobei bei unserem potentialtheoretischen Strömungsproblem unendlich hohe Geschwindigkeiten auftreten. Insbesondere tritt eine starke Umströmung der Ecken auf, wenn sich der
Querschnitt nur in einer Richtung ändert. Bei einer realen Strömung führt diese Kantenumströmung zu einer abgehenden Wirbel-
fläche, die sich zu einem Wirbel aufrollt.

Die Aufgabe besteht nun darin, solche Querschnittsänderungen oder anders ausgedrückt eine solche auf den einzelnen
Rechteckseiten konstante Normalkomponentenverteilung zu finden,
die keine Umströmung der Ecken ergibt. Die Aufgabe läßt sich
durch Anbringen einer geeigneten Quellbelegung auf den Seiten der
Querschnittsform und seines Spiegelbildes lösen, die man z.B. mit
Hilfe eines zweidimensionalen Pannelverfahrens finden kann.

Zur Vereinfachung unserer Rechnungen können wir das Problem noch etwas weiter idealisieren, ohne daß die wesentlichen
Strömungsvorgänge am oberen Teil des Fahrzeuges wesentlich beein-
flußt werden. Hierzu denken wir uns die Seitenflächen des Fahrzeuges bis auf den Erdboden heruntergezogen, d.h. wir betrachten
ein Fahrzeug mit Bodenabstand Null. Unsere vorher beschriebene
Potentialaufgabe reduziert sich dann auf das folgende Problem

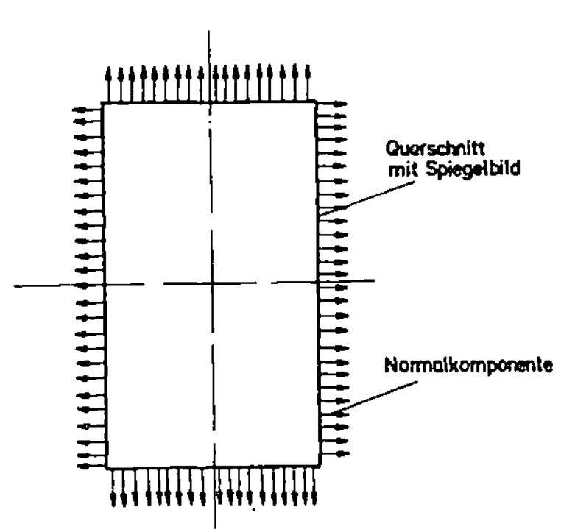

Bild 3: Weitere Idealisierung
der Normalkomponenten-
verteilung

(Bild 3). Wir haben ein Rechteck (bestehend aus Fahrzeugquerschnitt und Spiegelbild) auf dessen Ober- und Unterseite die Strömung mit der gleichen konstanten Normalgeschwindigkeit austritt und auf dessen Seitenflächen sie mit einer anderen konstanten Normalgeschwindigkeit austritt. Zu bestimmen ist das Verhältnis der beiden Normalgeschwindigkeiten, bei dem keine Eckenumströmung auftritt. Offensichtlich hängt dieses Geschwindigkeitsverhältnis nur vom Seitenverhältnis des Rechteckes ab. Zu seiner Berechnung kann man jetzt die Methode der konformen Abbildung benutzen, wobei das Rechteck auf eine Halbebene abgebildet wird und die Strömungsbedingungen in der Bildebene erfüllt werden. Auf diese Weise lassen sich nun in der durch die "slender-body" Theorie gegebenen Näherung Karosserieformen so auslegen, daß keine Umströmung der oberen Längskante des Fahrzeugs auftritt, bzw. solche mit gewünschter Intensität.

Um nun eine Grundform mit möglichst geringem Widerstand zu erhalten, müssen neben der eben geschilderten Bedingung noch einige weitere Bedingungen eingehalten werden. Insbesondere müssen die senkrecht zur Längsrichtung verlaufenden Kanten am Beginn des Kühlers und am oberen Rand der Front- und Heckscheibe so abgerundet sein, daß hier möglichst keine Ablösung auftritt. Ferner darf das Fahrzeugheck nicht zu steil eingezogen sein. Auf der anderen Seite muß man natürlich auch einige Bedingungen an die Grundform stellen, die es ermöglichen, sie ohne allzu große Änderungen als Kraftfahrzeugkarosserie verwenden zu können.

5.   Entwurf der Modelle

Im Hinblick auf eine einfache Versuchsdurchführung und Modellherstellung wurden die ersten Versuche zur Prüfung der dargestellten Überlegungen an einfachen Doppelmodellen (Grundform plus Spiegelbild) durchgeführt. Der Einfluß des Erdbodens wird hierbei durch die Symmetriebedingung simuliert. Die Modelle wurden so idealisiert, daß sämtliche Querschnitte Rechtecke mit abgerundeten Ecken sind. Um bei den Messungen eine gute Vergleichsmöglichkeit mit üblichen Karosserieformen zu erhalten, wurde auch eine Grundform einer normalen Stufenhecklimousine untersucht (die anderen Karosserieformen geben keine wesentlich besseren c_W-Werte). Gleichzeitig werden hierdurch bei dem Vergleich eventuelle Kanaleinflüsse und Re-Zahleinflüsse ausgeschaltet.

Bild 4 stellt das Doppelmodell der Grundform der normalen Stufenhecklimousine dar (Modell A). Die normalerweise vorhandene stumpfe Frontfläche (Kühlereinlauf) ist dabei durch Vorsetzen einer halbzylinderförmigen Kappe abgedeckt, um ein Abreißen der

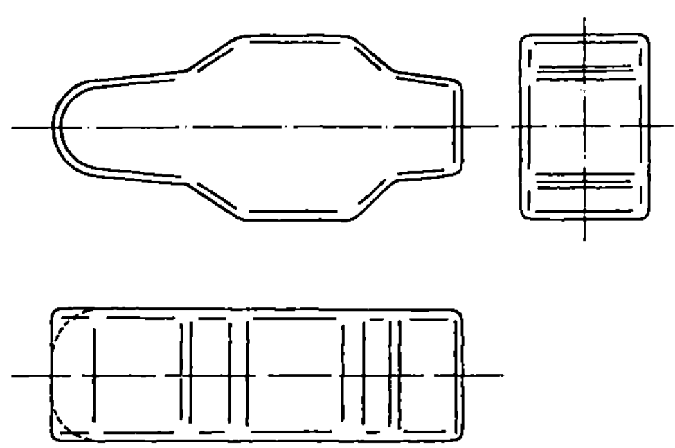

Bild 4: Idealisiertes Doppelmodell
einer Stufenhecklimousine
(Modell A bzw. A')

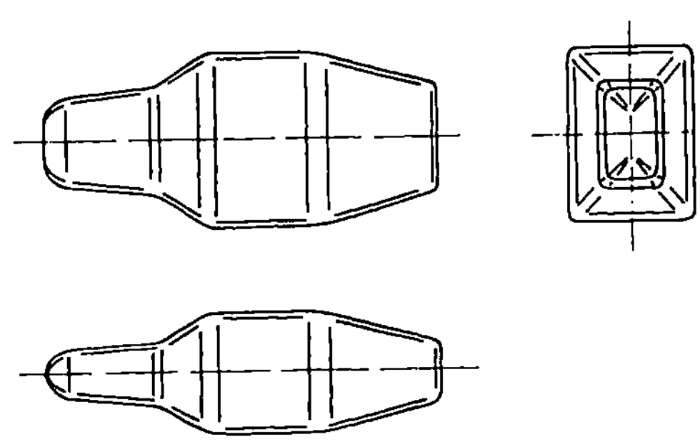

Bild 5: Idealisiertes Doppelmodell
ohne Längskantenumströmung
(Modell B)

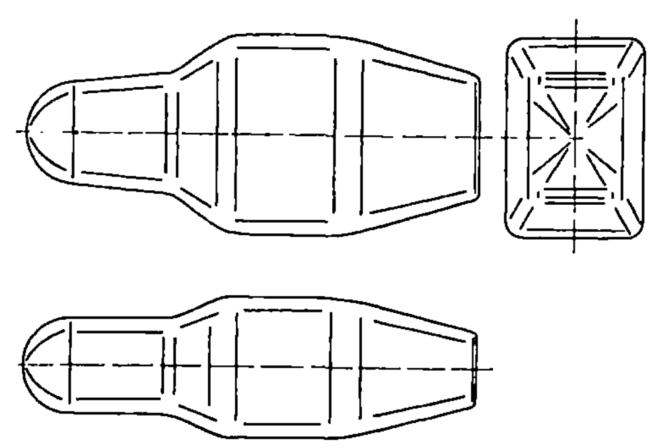

Bild 6: Idealisiertes Doppelmodell
ohne Ablösung an der oberen
Längskante des Fahrzeuges

Strömung wegen des nicht simulierten Kühlerdurchflusses zu vermeiden. Die Kanten dieser Abdeckung wurden mit dem gleichen Radius wie die anderen Kanten des Modells abgerundet. Von diesem Modell wurde noch eine zweite Variante untersucht (Modell A'), die sich vom Modell A nur in der Form der Kühlerabdeckung unterscheidet. Sie wurde auch an den Seiten mit dem Halbzylinderradius abgerundet (gestrichelte Linien im Grundriß von Bild 4).

Bild 5 stellt nun eine Grundform dar (Modell B), die so ausgelegt wurde, daß möglichst keine seitliche Umströmung der Längskanten stattfinden soll. Gleichzeitig wurde dabei die Bedingung eingehalten, daß das Modell praktisch die gleiche Fahrgastzelle (nach Form und Größe) wie Modell A besitzt, daß die Frontscheibe die gleiche Neigung und gleiche Höhe besitzt und die Länge annähernd so groß wie beim Modell A ist. Der Längskantenabrundungsradius wurde auch gleich groß wie bei Modell A gewählt. Dieses Modell sollte im wesentlichen dazu dienen, die Richtigkeit des gewählten Widerstandsreduzierungskonzeptes zu prüfen.

Bild 6 stellt eine weitere Grundform dar (Modell C). Sie besitzt das gleiche Heck, die gleiche Fahrgastzelle und die gleiche Frontscheibenhöhe und -neigung wie Modell B. Im vorderen Bereich wurde aber bewußt eine gewisse Längskantenumströmung zugelassen, die so bemessen wurde, daß noch keine Ablösung eintritt (soweit bei Fehlen von

Grenzschichttheorien für dreidimensionale Ablösung möglich). Auf
diese Weise konnte eine gegenüber Modell B wesentlich realisti-
schere Grundform erreicht werden. Der Längskantenabrundungsradius
wurde gegenüber Modell B zur Verringerung der Ablösegefahr etwas
erhöht.

6.   Meßergebnisse an den Doppelmodellen

An den vier Doppelmodellen A, A', B und C wurden Kraft-
messungen mit einer 6-Komponenten-Task-Waage im Hochgeschwindig-
keitskanal 75 x 75 cm^2 der DFVLR durchgeführt. Die Modelle waren
dabei mit einem zentralen Stiel von hinten gehalten. Die Messungen wurden bei zwei Geschwindigkeiten durchgeführt, die Mach-
zahlen von etwa 0,4 und 0,5 entsprachen. Bei einem Modellmaßstab
von 1:20 gegenüber einem Originalkraftfahrzeug wurden dabei Re-
Zahlen, bezogen auf die Fahrzeuglänge, von etwa 2,0 x 10^6 bzw.
2,4 x 10^6 erreicht. Der Einfluß der Machzahl bzw. der Re-Zahl
war dabei nur geringfügig, so daß nur die Messungen bei M = 0,5
dargestellt sind.

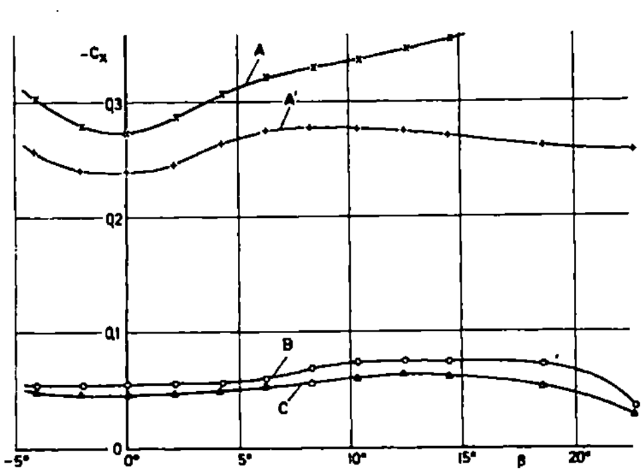

Bild 7 zeigt nun in
Abhängigkeit vom Schiebe-
winkel β den für den Fahr-
zeugwiderstand maßgebenden
Tangentialkraftbeiwert C_X
(bei β = 0 mit dem C_W-Wert
identisch) für die vier
Modelle A, A', B und C.
Betrachten wir nun zunächst
die Modelle A und A', die
ja der Grundform einer
Stufenhecklimousine ent-
sprechen, so erhält man C_W-
Werte von 0,27 bzw. 0,24.
Diese Werte liegen für eine
Grundform ohne Räder etwas
hoch, lassen sich aber dadurch erklären, daß bei dieser idealisierten Form so-
wohl Frontscheibe wie Seitenflächen und Dachfläche
keine Wölbung besitzen. Der
relativ große C_W-Unterschied

Bild 7: Tangentialkraftbeiwert C_X
(C_W-Wert) für die Modelle
A, A', B und C über dem
Schiebewinkel β
(Re_1 2,4 x 10^6)

bei den beiden Formen, die sich ja nur durch die Kühlerabdeckung
unterscheiden, kann durch eine räumliche Ablösung an der seitlichen Kante der Abdeckung beim Modell A erklärt werden.

Die Modelle B und C besitzen wesentlich kleinere C_W-Werte,
die in der Größenordnung von 1/5 der C_W-Werte der Modelle A und
A' liegen. Auffällig ist, daß der C_W-Wert beim Modell C, bei dem
ja eine gewisse Kantenumströmung zugelassen war, noch deutlich
tiefer liegt, als beim Modell B. Offensichtlich ist hier das Aus-
legungsziel, die räumliche Ablösung bei der Längskantenumströmung
zu vermeiden, voll erreicht, was auch durch Anstrichbilder bestätigt wurde. Der Grund dafür, daß der C_W-Wert beim Modell C
sogar noch tiefer liegt als beim Modell B, liegt wahrscheinlich

an den günstiger gewählten Abrundungsradien der Querkanten des
Modells. Die günstigen Widerstandseigenschaften bleiben auch bei
Seitenwind voll erhalten. Wie man aus Bild 7 erkennt, ist bei
kleinen Schiebewinkeln im Bereich von 5° der relative Gewinn sogar noch etwas größer als bei 0°, während der absolute Gewinn
sogar bei allen gemessenen Schiebewinkeln bis $\beta = 23°$ höher ist
als bei $\beta = 0°$. Bezüglich des Seitenkraftverhaltens sind die
Modelle B und C nicht ungünstiger als die Modelle A und A'. Die
Seitenkraft greift bei den ersteren zwar weiter vorn an, ist dafür aber kleiner, so daß sich für einen Bezugspunkt in Modell-
mitte etwa das gleiche Giermoment ergibt. Das Auftriebsverhalten
konnte bei der beschriebenen Versuchsanordnung natürlich nicht
gemessen werden.

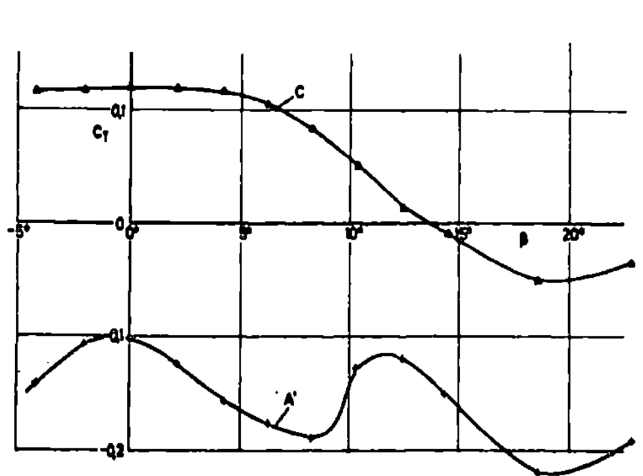

Bild 8 zeigt den eben-
falls mitgemessenen Tot-
wasserdruckbeiwert C_T für
die beiden Modelle A' und
C. Er liegt bei Modell C
um etwa 0,2 höher als bei
Modell A'. Man sieht hier-
aus, daß durch das Vermeiden der Ablösungen an den
Längskanten der Totwasserdruck (Druck an der stumpfen Heckfläche) günstig be-
einflußt wird.

Bild 8: Totwasserdruck C_T für die
Modelle A' und C

7. Meßergebnisse an Einzelmodellen mit Bodensimulation

Zur Prüfung der Frage, ob sich die erzielten guten Ergebnisse an den Doppelmodellen auch an Einzelmodellen bei der üblichen Simulation des Bodens durch eine Bodenplatte bestätigen,
wurden von den beiden Modellformen A' und C Einzelmodelle im Maß-
stab 1:4 für den 3m-Kanal der DFVLR gebaut (Modell A'_E und
Modell C_E).

Bild 9 zeigt das Modell A'_E. Gegenüber dem Doppelmodell A'
wurde hier nur die Kühlereinlaufabdeckung geändert und ein Heckdiffusor angebracht. Bild 10 zeigt das Modell C_E. Auch hier wurde
die Kühlereinlaufabdeckung in der gleichen Weise abgeändert und
ein Heckdiffusor angebracht. Zusätzlich wurde an diesem Modell
noch die aus Bild 10 ersichtliche Delle in der Bodenplatte angebracht, die die Gefahr einer Ablösung an der unteren Längskante
des Modells herabsetzen soll. Ferner wurden beide Modelle mit
abnehmbaren Radattrappen in Form von Scheibensegmenten ausgerüs-
tet. Von diesen unbedeutenden Änderungen abgesehen, waren die
Modelle A' und A'_E und die Modelle C und C_E geometrisch ähnlich.

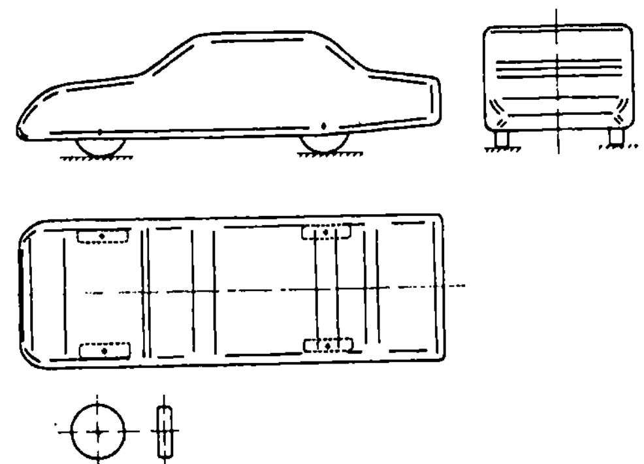

<u>Bild 9:</u>   Modell A' zugeordnetes
           Einzelmodell A'$_E$

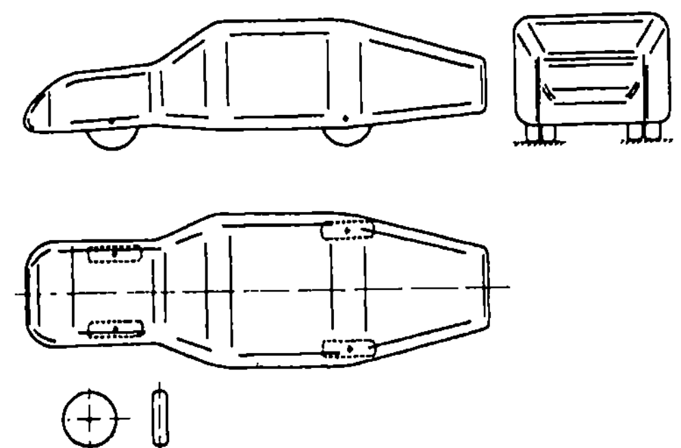

<u>Bild 10:</u> Modell C zugeordnetes
           Einzelmodell C$_E$

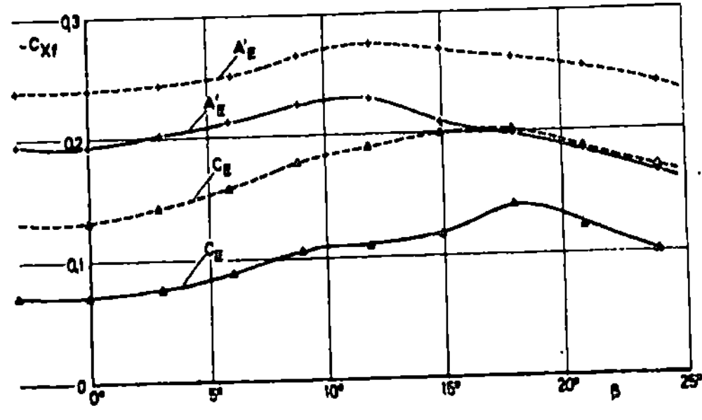

<u>Bild 11:</u> Tangentialkraftbeiwert C$_X$
           (C$_W$-Wert) für die Modelle
           A'$_E$ und C$_E$ über dem Schie-
           bewinkel β.
           ――― ohne Radattrappen,
           ----- mit Radattrappen

An beiden Modellen wurden dann für verschiedene Schiebewinkel 6-Komponentenkraftmessungen jeweils ohne und mit Radattrappen durchgeführt. Bei den Messungen ohne Radattrappen wurden die Modelle durch vier 8 mm starke, runde Bolzen gehalten. Der rechnerisch auf diese Bolzen entfallende Widerstandsanteil wurde als Korrektur abgezogen. Bei den Messungen mit Radattrappen waren diese Bolzen durch die Radattrappen abgedeckt. Als Bezugsfläche für die Kraft- und Momentenbeiwerte wurde in allen Fällen die Stirnfläche der Modelle ohne Radattrappen benutzt. Die Re-Zahl, bezogen auf die Modellänge, betrug etwa 4,8 x 10^6 und war damit etwa doppelt so hoch wie bei den Doppelmodellen.

<u>Bild 11</u> zeigt nun die gemessenen Tangentialkraftbeiwerte in Abhängigkeit vom Schiebewinkel β für die Modelle A'$_E$ und C$_E$, einmal ohne (ausgezogene Linien) und einmal mit Radattrappen (punktierte Linien). Betrachten wir zunächst den Beiwert für das Modell C$_E$ ohne Radattrappen. Hier ist der C$_W$-Wert gegenüber dem entsprechenden Doppelmodell C von C$_X$ = 0,046 auf C$_X$ = 0,070 gestiegen. Diese Steigerung war zu erwarten, da jetzt ja gegenüber dem Doppelmodell der Reibungswiderstand der Bodenplatte und eventuell entstehende kleine Verluste bei der Umströmung der unteren Längskante und eine eventuelle ungünstige Änderung des Totwasserdruckes hinzukommen. Beim C$_X$-Wert für das Modell A'$_E$ ist gegenüber dem Doppelmodell

eine Senkung des C_x-Wertes von 0,23 auf 0,19 eingetreten, obwohl auch hier der Reibungswiderstand an der Bodenplatte hinzukommt. Dieser Reibungsanteil wird aber sicher mehr als ausgeglichen durch die bekannte, günstige Wirkung des Heckdiffusors bei normalen Karosserieformen. Zusammenfassend läßt sich sagen, daß die günstigen Ergebnisse bei den Messungen an den Doppelmodellen auch an den Einzelmodellen voll bestätigt wurden. Die C_W-Wert Reduktion beim Übergang vom Modell A' auf C auf 20 % konnte bei Einzelmodellen natürlich nicht mehr erwartet werden. Immerhin trat aber auch hier beim Übergang von Modell A'_E auf C_E eine Senkung auf 37 % ein.

Betrachten wir nun die Ergebnisse an den Modellen mit Radattrappen, die durch die gestrichelten Linien in Bild 11 dargestellt sind. Hierbei wurde der Beiwert mit derselben Bezugsfläche gebildet, wie bei den Modellen ohne Radattrappe, d.h. die Stirnfläche der Räder wurde nicht mit zur Bezugsfläche gerechnet. Die Widerstandserhöhung durch die Räder ist bei Modell C_E höher als beim Modell A'_E. Dies ist zum Teil sicher darauf zurückzuführen, daß bei Modell A'_E Vorder- und Hinterräder die gleiche Spur besitzen, während beim Modell C_E die Vorderräder um mehr als eine Radbreite gegenüber den Hinterrädern versetzt angebracht sind, wodurch die Windschattenwirkung auf die Hinterräder beim Modell C_E geringer ist. Zum anderen ist das Modell C_E vermutlich empfindlicher auf Störungen der Strömung. Es bleibt eine Reduzierung des Widerstandes auf 55 % beim Übergang vom Modell A'_E auf C_E.

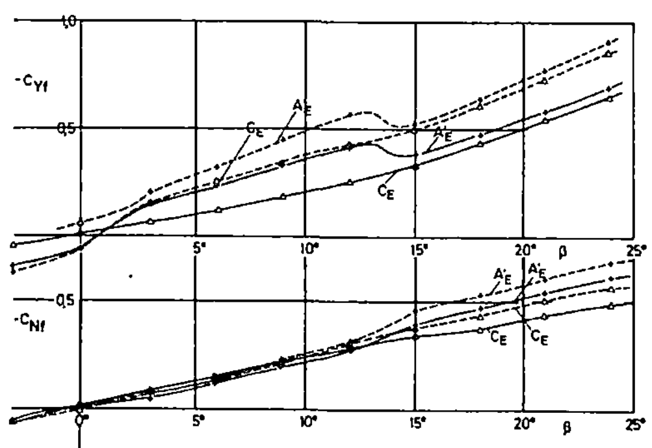

Bild 12: Seitenkraftbeiwert C_Y und Giermomentenbeiwert C_N für die Modelle A'_E und C_E. (——— ohne Radattrappe, – – – mit Radattrappe)

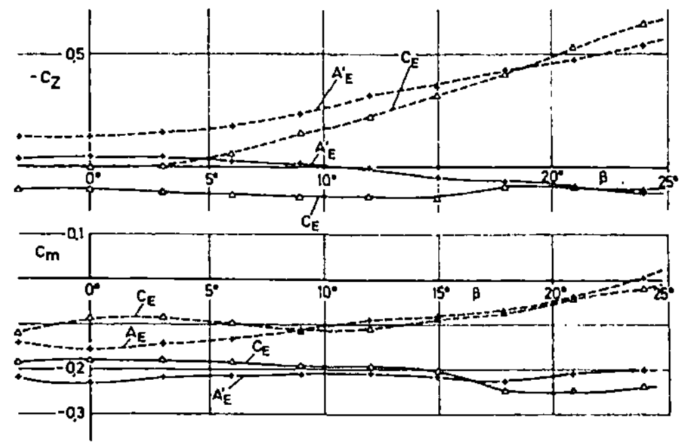

Bild 13: Normalkraftbeiwert C_Z (Auftriebsbeiwert) und Kippmomentenbeiwert C_M für Modelle A'_E und C_E. (——— ohne Radattrappe, – – – mit Radattrappe)

Bild 12 zeigt Seitenkraftbeiwert C_Y und Giermomentenbeiwert C_N für die beiden Modelle. Auch hier ist das C_E-Modell dem A'_E-Modell durchaus gleichwertig. Gleiches gilt für den in Bild 13 dargestellten Normalkraftbeiwert (Auftriebsbeiwert) und das zugehörige Moment.

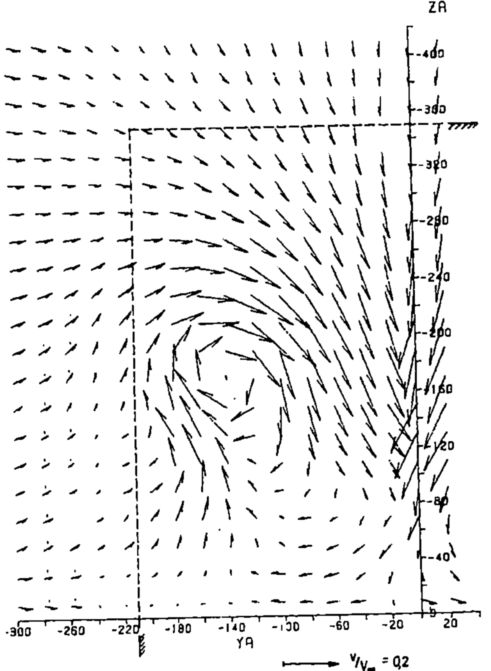

Bild 14: Querströmung in
einer Ebene,
400 mm hinter dem
Modell A'_E

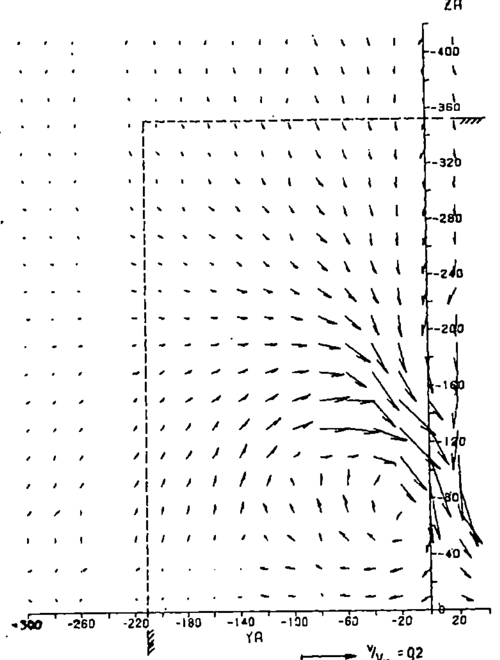

Bild 15: Querströmung in
einer Ebene,
400 mm hinter dem
Modell C_E

In $\underline{Bild\ 14}$ und $\underline{Bild\ 15}$ ist für die Modelle A'_E und C_E in einer Ebene, die 400 mm hinter dem Modell lag, die sogenannte Querströmung (in diese Ebene fallende Geschwindigkeitsvektoren) dargestellt, die mittels einer automatischen Richtungssonde gemessen wurde. Die y-Achse fällt hier mit dem Erdboden zusammen, während die z-Achse in die Symmetrieebene fällt. Gestrichelt ist die linke Hälfte des Fahrzeugquerschnittes eingezeichnet. Man erkennt beim Modell A'_E (Bild 14) deutlich den starken Längswirbel, der durch räumliche Ablösung an der oberen Längskante hervorgerufen wird. Auch beim Modell C_E (Bild 15) erkennt man einen Wirbel, der aber wesentlich schwächer und weniger ausgeprägt ist. Es sei an dieser Stelle darauf hingewiesen, daß auch beim Erreichen des Auslegungszieles (völlige Vermeidung der räumlichen Ablösung an der Längskante) in dem Totwassergebiet hinter einem nicht rotationssymmetrischen Modell Längswirbel zu erwarten sind. Bei diesen Wirbeln ist aber die zugeordnete Rotation (Komponente des Rotationsvektors normal zur Meßebene) im viel stärkeren Maße verschmiert und erstreckt sich praktisch über das gesamte Nachlaufgebiet. Diese Wirbel entstehen durch Drehen des direkt hinter dem Modell annähernd parallel zur Meßebene liegenden Rotationsvektors der Nachlaufströmung und benötigen für ihre Entstehung keinen zusätzlichen Energieaufwand.

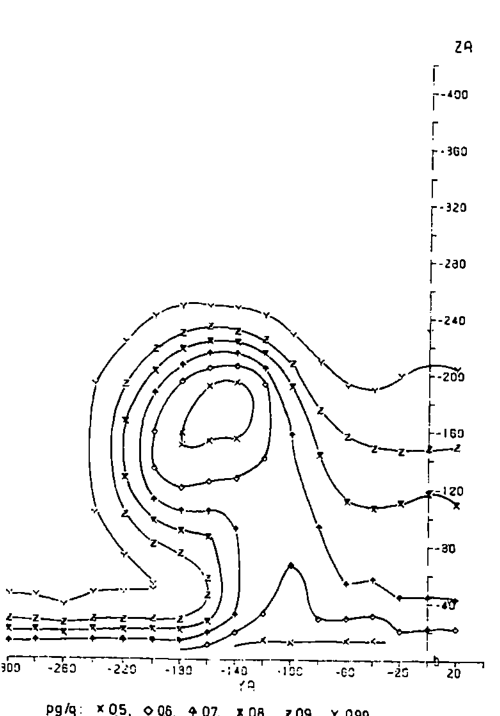

In viel deutlicherem Maße erkennt man dies aus den zugehörigen Gesamtdruckmessungen für die Modelle A'_E (Bild 16) und C_E (Bild 17). Hier ist beim Modell C_E der Längswirbel gar nicht mehr von der Nachlaufströmung unterscheidbar, während beim Modell A'_E der Längswirbel, der in seinem Kern sogar einen viel höheren Gesamtdruckverlust besitzt als die eigentliche Totwasserströmung, neben dem Nachlauf deutlich hervortritt.

$\underline{Bild\ 16}$: Gesamtdruckverlust in einer Ebene 400 mm hinter dem Modell A'_E

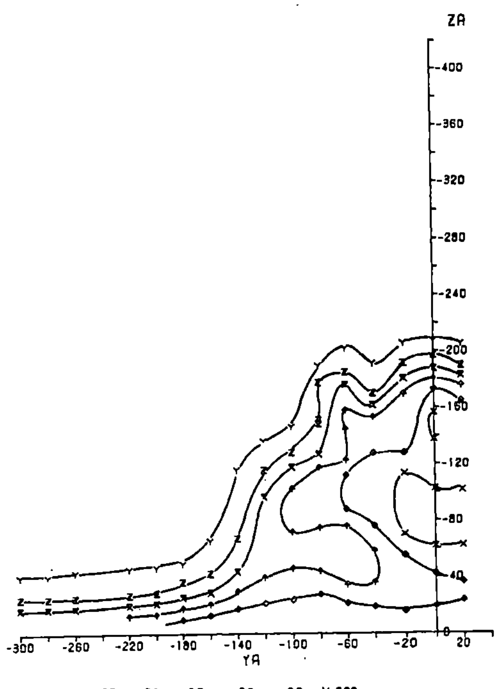

Bild 17: Gesamtdruckverlust in einer Ebene 400 mm hinter dem Modell C_E

8. Literatur

Buchheim, R., Deutenbach, K.-R., Lückoff, H.J.
Necessity and Premises for Reducing the Aerodynamic Drag of
Future Passenger Cars. International Congress and Exposition
Cobo Hall, Detroit, Michigan, Febr. 1981.

Ahmed, S.R.
Wake Structure of Typical Automobile Shapes. Transactions of
the ASME 162, Vol. 103, March 1981.

Ahmed, S.R.
Analytische Verfahren zur Aerodynamik von Straßenfahrzeugen.
DFVLR-Nachrichten Heft 21 (Juli 1977).

ANALOGIES BETWEEN OSCILLATION AND ROTATION OF BODIES INDUCED OR INFLUENCED BY VORTEX SHEDDING

by

Hans J. Lugt

David W. Taylor Naval Ship Research and Development Center
Bethesda, Maryland, USA

Abstract

Both rotation and oscillation are intrinsically periodic. Such periodicity may be induced on a body through vortex shedding; or, if a body is forced to rotate or oscillate, vortex shedding can strongly influence the periodic characteristics of the flow. This paper discusses various degrees of analogies or similarities between rotating and oscillating bodies with vortex shedding in a parallel flow, with respect to both the physical phenomena involved and the mathematical modelling. The latter is performed with the aid of the nonlinear oscillator concept or, more realistically, by solutions of the field equations. It is argued that, qualitatively, a fifth-order polynomial for the damping term is necessary in the oscillator model to simulate vortex-shedding effects. The feasibility of solving the Navier-Stokes equations for those complex flow phenomena is demonstrated.

1. Introduction

Two historical remarks of significance for understanding the general relationship between rotation and oscillation may introduce the reader to the subject of this paper.

(1) Aristotle [1] recognized that rotation and oscillation have in common that they are the only motions possible in a bounded and fixed space, and hence they must be periodic. However, only rotation can be steady. Aristotle used this concept of circular motion as perpetual and constant as the basis for a profound philosophical idea: He sought to establish with it the primacy of rotation over all other motions, exemplified by the eternal and unchanging revolution (so he thought) of the stars and planets.

Both the similar and different features of rotation and oscillation have a bearing on the process of vortex shedding from rotating and oscillating bodies. In a fluid at rest, physical similarity is not apparent, but in a parallel stream vortex shedding from rotating and oscillating bodies leads to clearly similar flow characteristics.

(2) The kinship between rotation and oscillation is even more clearly revealed in the second remark: In 1880 Greenhill [2] showed, on the basis of Kirchhoff's work, that a falling plate is described in the framework of potential-flow theory by the pendulum equation: $J\ddot{\alpha} - c \sin 2\alpha = 0$, where J is the moment of inertia, α the angle of attack, and c a constant. This equation permits, for the same restoring force, two kinds of motion: rotation and

oscillation. Although this unified description of rotating and
oscillating plates is appealing, it is unrealistic since it neglects vortex shedding (Fig. 1). In 1906 Riabouchinsky [3] showed
convincingly, when he rejected an earlier (1853) explanation by
Maxwell, that the explanation of the·autorotation of a falling
plate requires the inclusion of discontinuity sheets. The model
of a pendulum will be taken up later in this paper and will include nonlinear damping terms which simulate the effects of vortex shedding.

The role of vortex shedding in the understanding of rotating and oscillating bodies in a parallel stream (to which this
paper is restricted), and the analogies and similarities between
the flow characteristics of these types of motion, are the subject of this paper. In the following description, the movement of
bodies enforced by a motor must be distinguished from that induced by vortex shedding as shown in the example of a falling
plate (Fig. 1). The flow properties of forced and vortex-induced
or free motions can be quite different.

The frequency of the rotating or oscillating body is desig-
nated f_B, which in the case of a rotating body is related to the
angular velocity Ω by $\Omega = 2\pi f_B$. The frequency of vortex shedding
from a fixed body is f_N, and that from a rotating or oscillating
body is f_V. The dimensionless frequency is the Strouhal number
$St = fD/U$, where f is the frequency (either f_B, f_N, or f_V), D a
length scale (usually the diameter of a circular cylinder), and
U the constant velocity of the parallel flow (when the reference
frame is fixed to the translating body). The dimensionless angu-
·lar velocity is called the roll parameter $p = \Omega D/U$. The Reynolds
number is defined by $Re = UD/\nu$, where ν is the kinematic viscosity of the fluid.

Vortex-induced or vortex-influenced rotation and oscillation
of bodies in a parallel flow are widely found in technical applications and in nature [4]. The following examples are enumerated:
Rotation: Spin of aircraft, rotating finned missiles. Oscillation:
Galloping cables, oscillating buildings, flight of birds and insects, swimming of fish, falling leaflets.

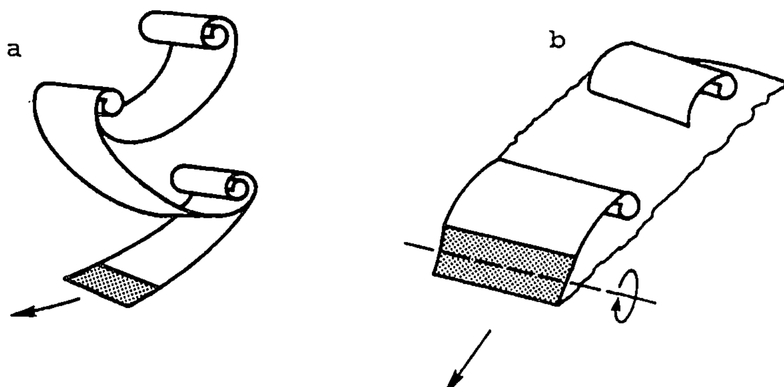

Figure 1: Vortex sheets rolling up periodically behind a plate
falling (a) in an oscillating and (b) in an autorotating way. In (b) the sheet behind the advancing edge
is irregular.

2. Steady and Quasi-steady Approach

The fluid flow produced by a rotating body in a parallel
stream (without shear) can be steady only if the body axis is parallel to the flow (as in the case of a propeller) or, when the
axis of rotation is perpendicular to the flow, if the body is a
circular cylinder or a sphere. In contrast, the flow around an
oscillating body is always unsteady. However, if the Strouhal number St_B is much smaller than unity, the flow around an oscillating body in a parallel stream can be considered "quasi-steady",
which means that the forces on the body vary so slowly with oscillation that they can be computed at a particular instant as if
the body were not oscillating. In this case a direct analogy
exists between rotation and oscillation in the sense that the modelling of the flow and its effect on the body are the same. This
similarity will be demonstrated with the phenomena of an autorotating Lanchester propeller and a galloping cylindrical body. Both
phenomena are nonlinear processes because of the existence of a
vortical wake.

Autorotation of the Lanchester propeller is explained by
means of Figs. 2 and 3. The propeller consists of a rectangular
blade which may be a thin board or may have a D-shaped cross-
section as in Fig. 2. The axis is parallel to the flow. Without
rotation, only a drag force D on a blade element is present (Fig.
3a). Slow rotation of the blade causes a lift L; but the resultant force F opposes rotation (b), so any slow initial rotation
ceases with time. However, if the initial rotation is large enough,
the resultant force can support rotation under certain conditions
(c). The propeller starts autorotating and increases its angular
velocity until a steady and stable equilibrium with the damping
force is reached.

The aerodynamic forces support rotation if $V/U = p > L/D$,
where V is the rotational velocity at the tip of the body. To
meet this criterion the angle of attack α must be in the region

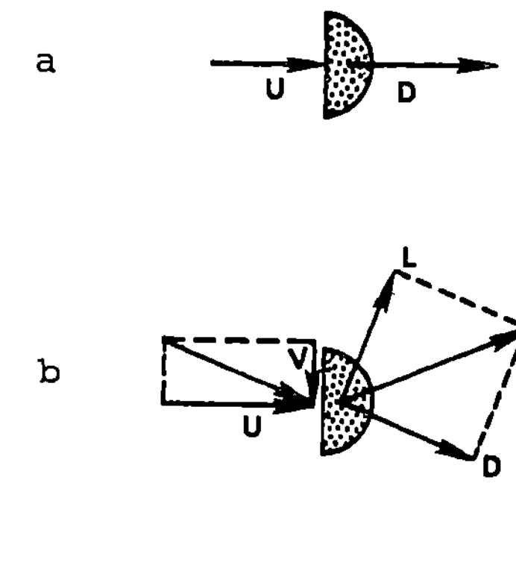

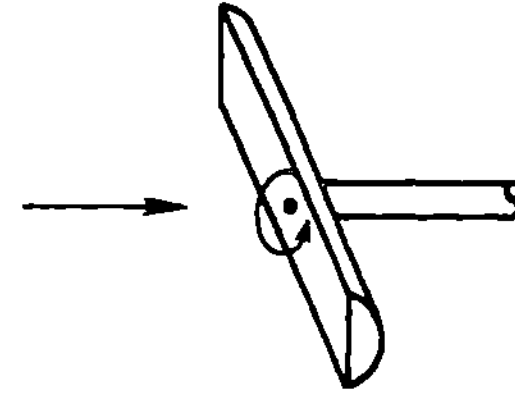

Figure 2:
Lanchester propeller or
Lanchester tourbillon.

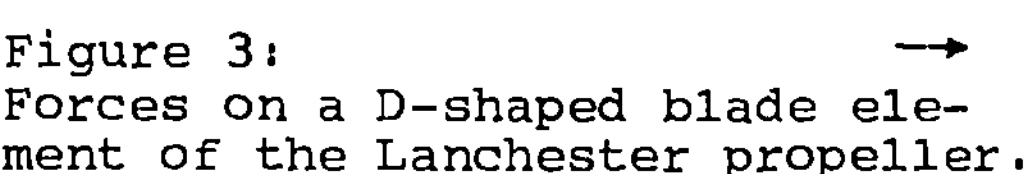

Figure 3:
Forces on a D-shaped blade element of the Lanchester propeller.

of stall so that the slope $dL/d\alpha$ of the lift curve becomes negative and its absolute value, according to Glauert [5], so large that the condition

$$\frac{dL}{d\alpha} + D < 0 \qquad (1)$$

is fulfilled.

The relation between initial impulse and final rate of autorotation may be illustrated more clearly by a method developed by Riabouchinsky [3]: A motor drives the propeller with constant angular velocity Ω , and the torque T acting on the propeller is measured as a function of Ω. The resulting curve is shown in Fig. 4. When Ω is relatively small or large, an outside positive torque is necessary to drive the propeller. Between these values an Ω-range can exist in which the outside torque is negative and thus has a braking effect on the rotation of the propeller. If the outside torque is removed, the propeller increases its angular velocity until T = 0, which is the state of autorotation (point A in Fig. 4). The other point at which T = 0, which is the initial value of Ω needed to induce autorotation, is unstable (point I in Fig. 4). The conditions for stable autorotation are thus

$$T = 0, \quad \frac{dT}{d\Omega} > 0 \qquad (2)$$

Qualitatively, the Riabouchinsky curve can be expressed by a fifth-order polynomial, which for rotation is

$$T/J = A\Omega + B\Omega^3 + C\Omega^5. \qquad (3)$$

The curves in Fig. 4 do not necessarily cross into the half space $T \leq 0$; that is, not all wings autorotate. Various blade profiles were investigated with respect to their side force by Parkinson and Brooks [6], among others, and are recorded in [7]. Of course, the local condition (1) for a blade element need not be met everywhere along the span as long as the integration over the whole span (for instance with a strip theory) results in a torque in the direction of rotation.

Den Hartog [8] explained, in an analogous way, the phenomenon of galloping, which is a special kind of body oscillation perpendicular to the parallel flow, characterized by $St_B < 0.1$.

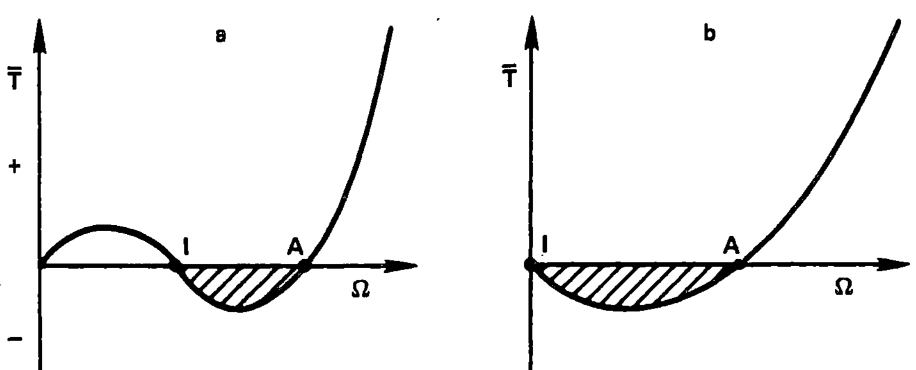

Figure 4: Riabouchinsky curves for (a) $\Omega_I > 0$, (b) $\Omega_I = 0$.

Again, when strip theory is applied (this time to describe a quasi-steady state), criterion (1) serves as a condition for self-excited and self-maintained oscillation. If the oscillation is enforced, the force acting on the body yields a Riabouchinsky curve with f_B as the abscissa. The physical reason, of course, is the same as for rotation: the existence of a vortical wake, resulting in $dL/d\alpha < 0$.

The analogy between rotation and oscillation in these situations is almost exact, except that rotation is truly steady (provided the wake is steady). Den Hartog [8] actually proposed to study blade profiles of galloping bodies by means of the Lanchester propeller.

A similar analogy exists between the rotating wings of aircraft and torsional oscillating wings, when one side of the wing stalls. Here too, steady and quasi-steady approaches, respectively, can be applied. The various forms of the Riabouchinsky curve are illustrated in graphs by Glauert [9].

This section has indicated that the Riabouchinsky curve is typical for forces or torques acting on bodies with vortical wakes. This observation will be confirmed by other examples of a quite different physical nature in the next section.

3. Fixed Body Axes Perpendicular to the Flow

There is no reference frame in which noncircular cylinders rotating with constant Ω perpendicular to a parallel flow cause steady flows. These flows are periodic, with vortex shedding over a wide range of Reynolds numbers. It is here that interesting similarities with bodies oscillating perpendicular to a parallel flow occur.

Consider a plate, its axis fixed in a parallel stream, rotating with constant angular velocity Ω maintained by a motor. The flow characteristics of this "forced rotation" depend primarily on p and to a lesser extent on Re (sufficiently greater than Re = 0). The torque averaged over one cycle and plotted versus p results in a Riabouchinsky curve (Fig. 4a).

If the plate rotates very rapidly compared to the parallel flow, that is, if p is larger than unity, vortex shedding takes place behind the edges of the plate when the reference frame rotates with the plate (Fig. 5a, small vortex at the lower edge). In this case the frequency of vortex shedding may differ from the

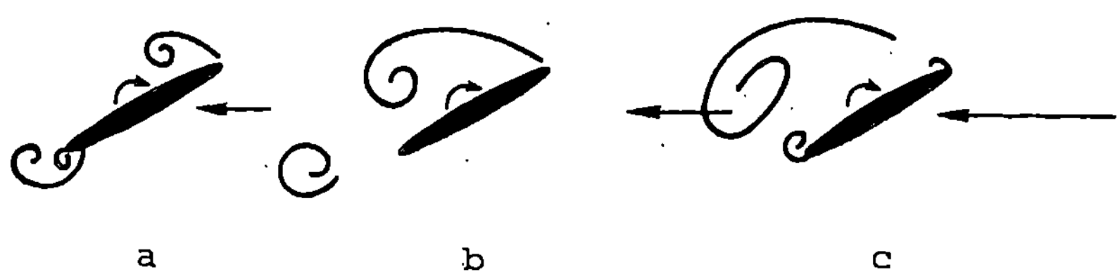

a   b   c

Figure 5: Sketch of vortex shedding about a power-driven rotating plate. (a) p > 1, (b) p ≈ 1, and (c) p ≪ 1 [9].

rate of rotation by integral values, showing the existence of superharmonic modes $\Omega = 2\pi(n+1)f_V$, $n = 1, 2, \ldots$ which have a simple physical explanation: a rapidly rotating plate traps vortices before it releases them after several revolutions.

Around $p = 1$ vortex shedding adjusts to the rate of rotation, a phenomenon called "synchronization" or "lock-in" (Fig. 5b): $2\pi f_V = \Omega$. In this range energy can be transferred from the fluid flow to the body (shaded region in Fig. 4a).

If p is much smaller than unity, even approaching $p = 0$, (that is, the subharmonic region) the frequency of vortex shedding is greater than the rate of rotation, $\Omega < 2\pi f_V$, and becomes independent of Ω for $p \to 0$ (Fig. 5c). Vortices then separate behind the edges as seen from the upstream parallel flow.

With the transfer of energy from the flow to the body in the "lock-in" range, a freely rotating plate may reach self-sustained spinning, which is another kind of autorotation.

The explanation of autorotation is quite involved, and only the major aspect is mentioned here: at the critical angle of attack ($\alpha \approx 135°$), which determines autorotation and is shown in Fig. 5, a value of p exists at which the flow passes over the rear end of the plate without creating a vortex. Calculation of the pressure distribution reveals that this situation supports autorotation. Details are given in [10].

Below Re ≈ 1000, autorotation becomes Re-dependent, for instance, $p = 0.5$ at Re $= 200$; and autorotation ceases when Re $\to 0$.

Flow characteristics analogous to those of forced rotation are found for forced oscillation. Here, two types may be distinguished, depending on whether the body oscillates about a fixed axis or translates with its axis sideways to the flow. Since the latter case has been investigated more extensively, this discussion will focus on that type.

If a cylindrical body is forced to translational oscillations normal to a parallel stream, the average absolute value of the force required for one period, plotted against the frequency, will result again in a curve like that of Fig. 4. Near the natural frequency, lock-in occurs, and the force supports oscillation that would cause an amplitude increase if the body were free to move. This phenomenon, analogous to autorotation of plates, can be explained by the lift force which acts on the body in a certain rhythm related to vortex shedding. This lift force, when synchronized with the forced oscillation, can be larger than that required for the forced oscillation [11].

The source of the lift force can be found in the shift of the separation points for circular cylinders, and in the change of angle of attack for rectangular cylinders [12].

Lock-in has also been observed for certain super- and subharmonic modes. This condition has been explained by Ericsson [13] as the effect of negative Magnus lift of rotating circular cylinders.

As in the case of rotating bodies, the dependence of the flow characteristics on the Reynolds number is small. The reason for this is - and this is true for both rotation and oscillation - that the flow field changes so fast that diffusion of vorticity

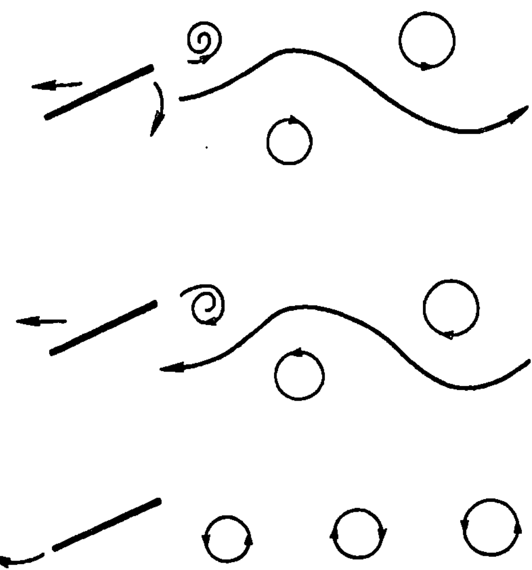

Figure 6: The three types of vortex shedding from forced oscillating plates which move to the left.

has not sufficient time to be of significance compared to inertial forces.

Forced torsional oscillation about a fixed axis in a parallel flow is of great importance in nature: It provides a means
for locomotion of fish, birds, and insects. As with the rotating
plate (Fig. 5) three cases may be distinguished, depending on the
way the vortices are shed from the plate's edges (Fig. 6): (1)
Vortex shedding from slowly oscillating plates; the fluid flow
exerts a drag on the body. (2) Synchronization of vortex shedding
and body oscillation; the body movement creates either a thrust
or a drag depending to a large extent on the location of the axis;
at Re ≈ 0 no thrust is produced. (3) Fast oscillation; again the
flow exerts a drag on the body [14].

In nature and technology a translational motion is often
superposed on the purely rotational oscillation for higher effi-
ciency; and in many examples the plate is not rigid but elastic
as in the case of fins of fish [15]. Change of angle of attack
combined with translational motion is used in the Weis-Fogh mechanism of certain insects [16]. The counterpart of rotating bodies is the Voith-Schneider propeller and, in the special case of
fixed blades, the Darrieus turbine [17].

Mathematical models which simulate vortex shedding from rotating and oscillating bodies may be divided into two categories:
(1) mechanical oscillator models consisting of one or two nonlinear ordinary differential equations, and (2) solutions of the
field equations, either the Navier-Stokes equations or truncated
versions of them.

(1) This paper is not the place to review the numerous oscillator models published [11]. Rather the physical processes discussed may be illustrated by solutions of a new ordinary differential equation which combines the overall characteristics of rota-

ting and oscillating bodies in the presence of vortex shedding.
The equation for the simple pendulum mentioned in the introduction is extended to include the vortical effects represented by
the fifth-order polynomial, that is, equation (3) of the Riabouchinsky curve:

$$\ddot{\alpha} + A\dot{\alpha} + B\dot{\alpha}^3 + C\dot{\alpha}^5 + D \sin 2\alpha = 0, \tag{4}$$

$$d\alpha/dt = \dot{\alpha} = \Omega, \tag{5}$$

where the coefficients A through D are constants. The initial con-
ditions are $\alpha(t_0) = \alpha_0$ and $\dot{\alpha}(t_0) = \dot{\alpha}_0 = \Omega_0$. (Landl [18], by the
way, used a nonlinear damping of the form $\alpha^4 \dot{\alpha}$). Figure 7 shows
the results of a parametric study in which values for $\dot{\alpha}$ and coefficients A through D were chosen which are close to the instability point Ω_I of Fig. 4a for a relatively large moment of inertia.
If $\dot{\alpha}_0$ is smaller than $\dot{\alpha}_I = \Omega_I$ the rotating plate will come to a
standstill with a few oscillations in the end phase (lower two
curves). If $\dot{\alpha}_0$ is larger than $\dot{\alpha}_I$, the plate will increase its angular velocity until it reaches the state of autorotation A,
which may be considered an "open limit cycle" (upper two curves).

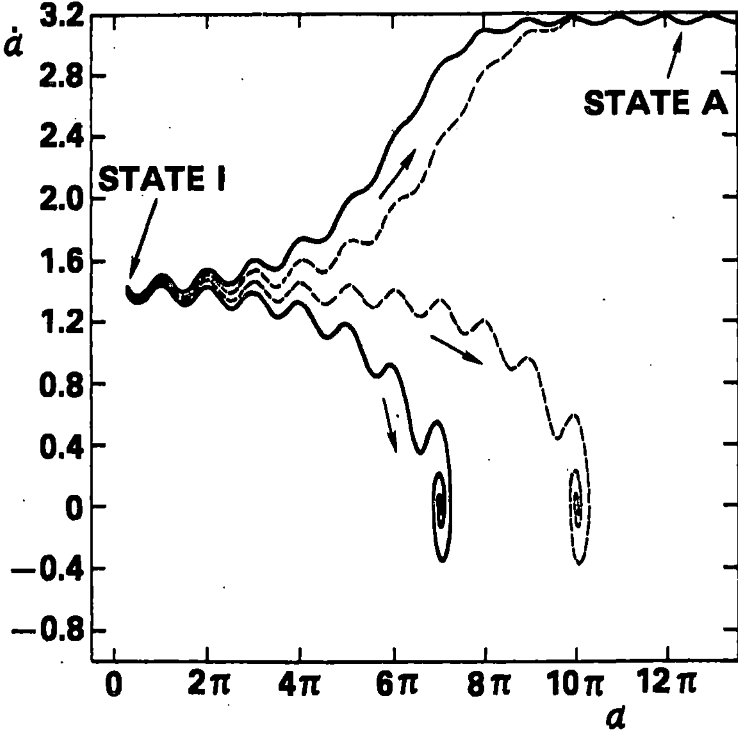

Figure 7: Phase plane of the extended pendulum equation (4) for
A = 0.2, B = 0.12, C = 0.01, D = 0.2, $\alpha_0 = \pi/4$, $\dot{\alpha}_0 =$
1.41, 1.42, 1.43, 1.44. The lowest value of $\dot{\alpha}_0$ corresponds to the lowest curve, etc. The time interval of
the Runge-Kutta method is $\Delta t = 0.1$.

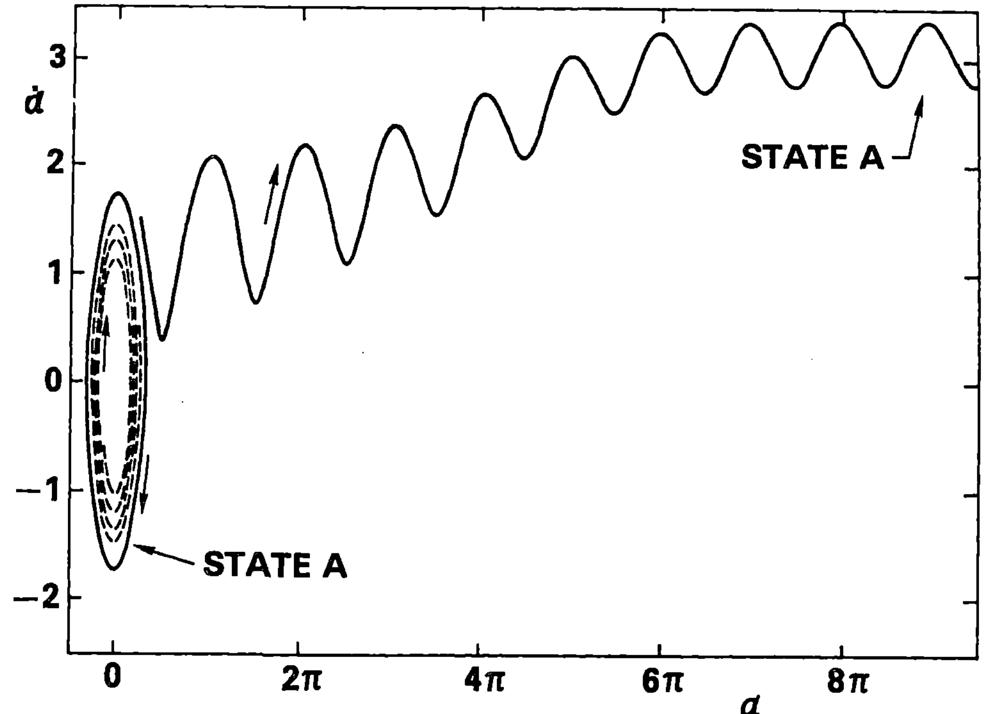

Figure 8: Phase plane for the same coefficients as in Fig. 7 but with D = 2, $\dot{\alpha}_O$ = 0.5, 1.0, 1.5, Δt = 0.05.

The point of instability is thus confined to the interval 1.42 < $\dot{\alpha}_I$ < 1.43. In Fig. 8 the transient periods in the phase plane with the same constant coefficients but with smaller moment of inertia (D = 2) and different $\dot{\alpha}_O$ are shown. Again, the lowest value of $\dot{\alpha}_O$ lies below $\dot{\alpha}_I$ and the movement of the body comes to rest but now in an oscillating way (curve represented by dashes). $\dot{\alpha}_O$ = 1.0 is above $\dot{\alpha}_I$ and is very close to self-sustained oscillation, illustrated by the "closed" limit cycle (which is considered a state A). At the highest value $\dot{\alpha}_O$ = 1.5 the plate rotates and increases its angular velocity to stable autorotation A ("open" limit cycle). The solutions in Fig. 8 reveal, therefore, two limit cycles of different kinds, whose existence depends on the initial push $\dot{\alpha}_O$.

Although the coefficients A through D can be approximated from experimental data, the pendulum model is not a true approximation of the basic field equations in the sense of a perturbation theory.

(2) Thus, ultimately, solutions of the Navier-Stokes equations or of their truncated versions must be sought. This approach has been successfully pursued for a number of problems. Figs. 9 and 10 show numerical solutions of the Navier-Stokes equations for an instant in torsional oscillation of an airfoil at Re = 1000, by Mehta [19], and for an incident of Weis-Fogh flapping, by Haussling [20]. An instantaneous picture of an autorotating plate, developed by the author, is given in Fig. 11 [10] and can be compared with a photo of a wind-tunnel experiment (Fig. 12). Sarpkaya

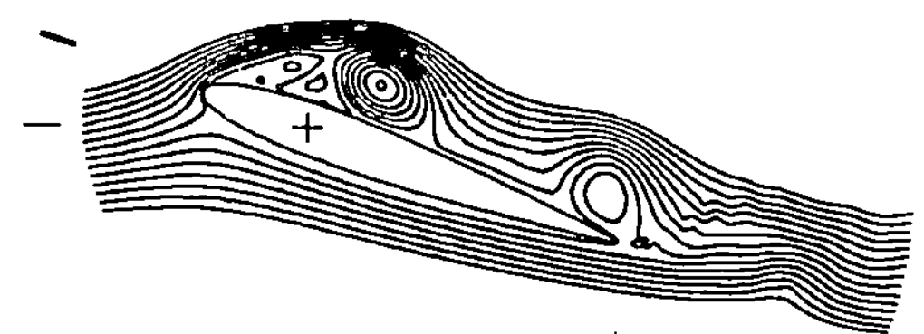

Figure 9: Instant in the torsional oscillation of an airfoil in
a parallel flow for Re = 1000 (computer-generated
streamlines by U.B. Mehta, NASA-AMES).

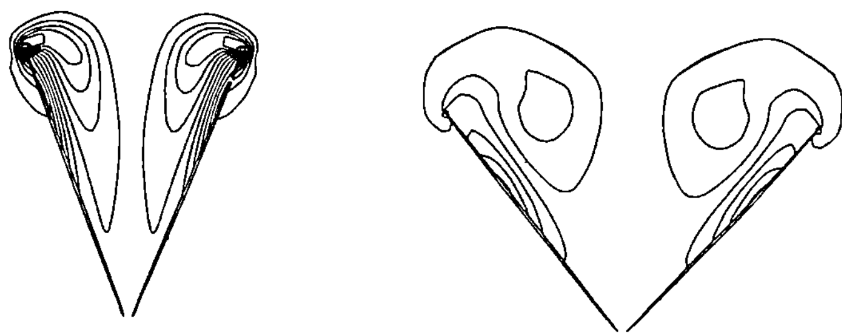

Figure 10: Vortex shedding during lift generation based on the
Weis-Fogh mechanism for Re = 30 (computer-generated
equi-vorticity lines by H.J. Haussling, DTNSRDC).

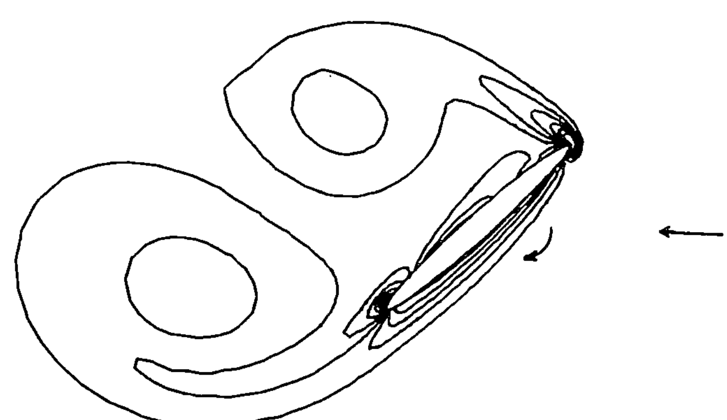

Figure 11: Vortex shedding behind the retreating edge of an autorotating thin elliptic cylinder at the critical angle
of attack 135° for Re = 200, p = 0.25 (computer-generated equi-vorticity lines by the author).

and Shoaff [21] also performed an interesting study using a point-
vortex method for a circular cylinder oscillating perpendicular
to the flow and were able to describe the typical flow characteristics of lock-in and amplitude increase.

Figure 12: Photo of vortex shedding behind the retreating edge
of an autorotating plate at about 140° for Re ≈ 10^5
and p ≈ 1 (from F.N.M. Brown, University of Notre
Dame, Indiana, USA).

4. Freely Moving Axes Perpendicular to the Flow

The axes of oscillating and rotating bodies do not have to
be fixed or forced on a prescribed-path but may move freely. Fig.
1 illustrates this for the falling plate. The overall flow field
in this case is more complex than when the axes are fixed, and at
present no solutions of the field equations exist in the literature.

Vortex shedding is affected by the movement of the body and
vice versa. Autorotation can even be explained by only the movement of the body in a quasi-steady way as was done by Maxwell
[22]. Another illustration of self-sustained rotation and oscillation without any frictional effects is Greenhill's potential-flow
solution, which results in the simple pendulum equation mentioned
in the introduction. However, exclusion of vortex shedding is unrealistic as was pointed out by Riabouchinsky.

For the freely falling plate the point of instability in
Fig. 4a can shift to the origin as indicated in Fig. 4b. This
means that the slightest disturbance causes the plate to fall
either in the autorotating or in the wobbling mode as shown in
Fig. 1. Two solutions of the extended pendulum equation (4) which
illustrate these cases are displayed in Figs. 13 and 14 in the
phase plane. Fig. 13 shows how the plate starts autorotating
after a brief wobbling period. Fig. 14 demonstrates the transition through increased wobbling to self-sustained oscillation.

The flow characteristics discussed so far are not restricted to plates. Falling circular disks exhibit the same behavior,
although the vortex patterns in the wake are more complex [23].
Fig.15 is a photo showing the wake of a falling oscillating disk.

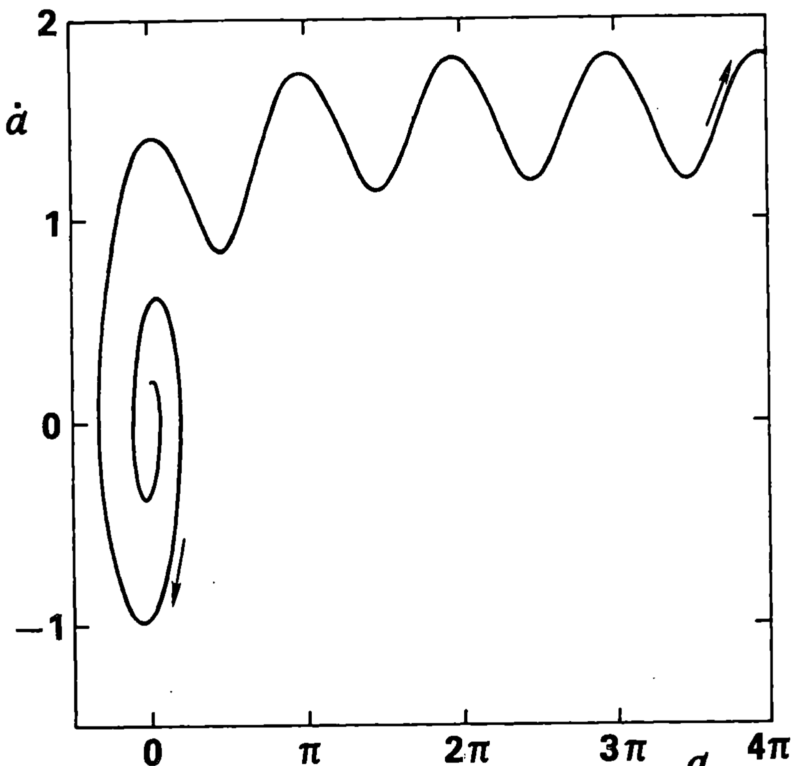

Figure 13: Phase plane of the extended pendulum equation (4) for
A = -0.5, B = -0.2, C = 0, D = 1.0, $\alpha_0 = 0^\circ$, $\dot{\alpha}_0 = 0.2$,
$\Delta t = 0.1$.

Self-sustained oscillation and rotation occur for Re > 100, and
the moment of inertia determines the mode of motion. In the range
1 < Re < 100 initial oscillations damp out to a steady fall in a
horizontal position. In this situation, point Ω_I does not coincide with the origin (Fig. 4a). Finally, in the slow-motion regime Re < 1, the falling disk remains in its initial orientation,
no matter how oblique that may be.

Nonlinear behavior due to vortex shedding can also be found
in other fluid dynamics problems. Of practical interest is the
motion of finned missiles in the so-called "roll speed-up" range
of angles of attack above ca. 36°. An evaluation of experimental
data by Cohen et al. [24] revealed a cubic relation between $\overline{T}$
(averaged torque) and $\dot{\alpha}$, complicated by the consideration of can-
ted fins. Other examples are rolling aircraft, particularly those
with delta wings which have leading-edge vortices, and the rolling motion of ships, in which bilge vortices are involved [25].
In these cases the flow is parallel to the axis of rotation or
oscillation, and not much is known - to the author's knowledge -
about the importance of higher nonlinear damping terms.

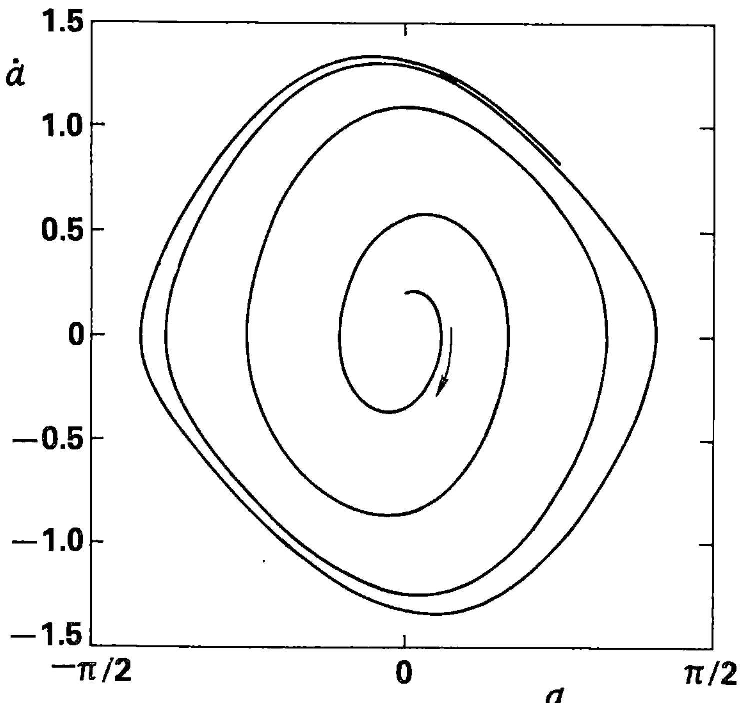

Figure 14: Phase plane of the extended pendulum equation (4) for
A = -0.5, B = -0.4, C = 0, D = 1.0, $\alpha_O = 0^O$, $\dot{\alpha}_O = 0.2$,
$\Delta t = 0.1$.

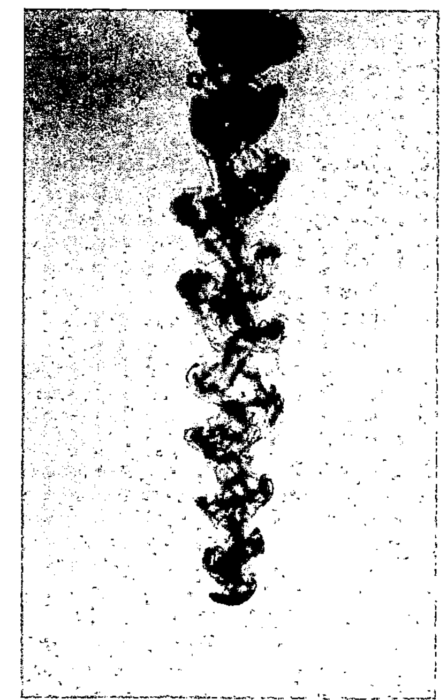

Figure 15:
Wake of a disk wobbling
down during fall. Re $\approx$
1500 (Photo by W.W.
Willmarth, University of
Michigan).

94

5. Concluding Remarks

The periodicity of the rotation and oscillation of bodies results in analogous or similar fluid motion. This is particularly true for flows around bodies with vortex shedding. Typical nonlinear properties due to vortex shedding are shared by both rotating and oscillating bodies in a parallel flow and include the lock-in effect, energy transfer from fluid flow to body, and existence of sub- and superharmonic modes. The Re-dependence beyond, say Re = 1000, is small.

Qualitatively, vortex-induced or vortex-influenced rotation and oscillation can be described by a fifth-order polynomial for the damping terms in the pendulum equation (Riabouchinsky curve, Fig. 4). It is suggested that this fundamental curve, verified experimentally and/or numerically for the Lanchester propeller, the autorotating plate, galloping, and vortex-induced vibration, is also applicable to other vortex-related phenomena like the autorotation-in-roll of aircraft and the rolling motion of ships.

Ultimately, the Navier-Stokes equations or their truncated versions must be solved numerically (with inclusion of turbulence models) to describe rotating and oscillating bodies in a real fluid. Initial effort in this direction for the laminar-flow range has been successful and is very promising.

Acknowledgment

Part of this study was performed while the author was a Humboldt fellow (Senior US Scientist Award) at the University of Karlsruhe, Institut für Strömungslehre und Strömungsmaschinen, Prof. Dr.-Ing. J. Zierep.

References

[1] Aristotle's Physics, transl. by R. Hope. University of Nebraska Press, Lincoln, 1961, 167.

[2] Lamb, H., Hydrodynamics. Sixth ed., Dover, New York, 1945, 174.

[3] Riabouchinsky, D.P., Thirty years of theoretical and experimental research in fluid mechanics. J. Roy. Aer. Soc. 39 (1935), 282 and 377.

[4] Lugt, H.J., Vortex Flow in Nature and Technology. In German: Braun, Karlsruhe, 1979. In English: to be published by Wiley Interscience, New York, in spring 1983.

[5] Glauert, H., The rotation of an aerofoil about a fixed axis. Advisory Committee for Aeronautics, Rep. & Memo. No. 595, March 1919, 443.

[6] Parkinson, G.V. and Brooks, N.P.H., On the Aeroelastic Instability of Bluff Cylinders. J. Appl. Mech., June 1961, 252.

[7] Blevins, R.D., Flow-induced vibration. Van Nostrand Reinhold Co., 1977.

[8] Den Hartog, J.P., Transmission Line Vibration Due to Sleet. Trans. AIEE 51 (1932), 1074.

[9] Lugt, H.J., Autorotation. Ann. Rev. Fluid Mech., Ann. Reviews, Inc., Palo Alto, California, Vol. 15, 1983.

[10] Lugt, H.J., Autorotation of an elliptic cylinder about an axis perpendicular to the flow. J. Fluid Mech. 99 (1980), 817.

[11] Sarpkaya, T., Vortex-Induced Oscillations . J. Appl. Mech. 46 (1979), 241.

[12] Nakamura, Y. and Mizota, T., Unsteady Lifts and Wakes of Oscillating Rectangular Prisms. J. Eng. Mech. Div., Dec. 1975, 855.

[13] Ericsson, L.E., Kármán Vortex Shedding and the Effect of Body Motion. AIAA J. 18 (1980), 935.

[14] Hertel, H., Structure-Form-Movement. Reinhold Publ. Co., 1966.

[15] Katz, J. and Weihs, D., Hydrodynamic propulsion by large amplitude oscillation of an airfoil with chordwise flexibility. J. Fluid Mech. 88 (1978), 485.

[16] Maxworthy, T., The Fluid Dynamics of Insect Flight. Ann. Rev. Fluid Mech., Vol. 13, 1981, 329.

[17] Strickland, J.H., Webster, B.T., and Nguyen, T., A Vortex Model of the Darrieus Turbine: An Analytical and Experimental Study. J. Fluid Eng. 101 (1979), 500.

[18] Landl, R., Theoretical model for vortex-excited oscillations. Intern. Symp. Vibration Problems in Industry. Keswick, England, 1973.

[19] Mehta, U.B., Dynamic Stall of an Oscillating Airfoil. AGARD CP-227, paper no. 23.

[20] Haussling, H.J., Boundary-Fitted Coordinates for Accurate Numerical Solution of Multibody Flow Problems. J. Comp. Phys. 30 (1979), 107.

[21] Sarpkaya, T. and Shoaff, R.L., A Discrete Vortex Analysis of Flow About Stationary and Transversely Oscillating Circular Cylinders. Tech. Rep. No. NPS-69SL79011, Naval Postgraduate School, Monterey, California, 1979.

[22] Maxwell, J.C., On a particular case of the descent of a heavy body in a resisting medium. Scient. Papers. Cambridge University Press, 1890, 115.

[23] Willmarth, W.W., Hawk, N.E., and Harvey, R.L., Steady and Unsteady Motions and Wakes of Freely Falling Disks. Phys. Fluids 7 (1964), 197.

[24] Cohen, C.J., Clare, T.A., and Stevens, F.L., Analysis of the Nonlinear Rolling Motion of Finned Missiles. AIAA J. 12 (1974), 303.

[25] Sachs, G. und Fohrer, W. Einfluß der aerodynamischen Kopplung auf die Flugzeugdynamik bei schnellen Rollbewegungen. Z. Flugw. Weltraumf. 4 (1980), 379.

A POINT VORTEX METHOD APPLIED TO INTERFACIAL WAVES

D.W. Moore
Imperial College
London

Abstract

The point vortex method can be adapted to apply to the evolution of waves on the interface between homogeneous incompressible fluids by adding equations governing the baroclinic change of point vortex strengths. It is shown that, in common with Lagrangian methods of other type, the method can fail through the growth of an instability. It is suggested that this instability, which causes waves on the scale of the inter-vortex separation to grow, is numerical and due to the distortion of the dispersion equation inherent in the replacement of the interfacial vortex sheet by an array of point vortices.

§1.　Introduction

Recent computational work on nonlinear interfacial motion in two dimensions has employed Lagrangian methods. The initial interface is defined by the coordinates of N marker particles. Provided that the equations of motion can be manipulated in such a way that they predict the velocity of each particle, given the positions of all, the motion of the marker particles and hence the evolution of the interface can be calculated. The best known Lagrangian method is that of Longuet-Higgins and Cokelet (1976).

A distinct class of Lagrangian methods is based on the recognition that the interface is a vortex sheet whose vorticity is generated baroclinically. The marker particles are point vortices and to the 2N ordinary differential equations describing the motion of the point vortices are added N equations for their strengths. Unfortunately, the rate of change of the strengths is determined by the pressure gradient along the interface (equation (2.4) below) and this in turn involves the rate of change of strengths through the fluid acceleration. Thus the rate of change of the strengths is determined implicitly and iteration must be used.

The point-vortex method has been used by Zarodny and Greenberg (1973), Kulyaev (1976), Baker, Meiron and Orszag (1980), Pullin (1982) and a version of the method is described in §2. Users of Lagrangian methods have noticed that attempts to follow the Stokes wave eventually fail. For example, Longuet-Higgins and Cokelet (l.c) reported that a zig-zag pattern appeared after a wave of amplitude of about 90% of that of the wave of greatest height had travelled about one-half a wavelength. Similar difficulties have been encountered by Forberg (private communication) who used a fast conformal mapping technique, Fornberg (1980), and the present work is aimed at understanding this phenomenon.

The numerical scheme of §2 is shown analytically in §3 to be stable when applied to a Stokes wave of infinitesimal amplitude. This is confirmed in §4, where it is shown computationally that the breakdown is caused by the growth of short waves with a growth rate roughly proportional to the wave amplitude. The analysis of §3 reveals that short waves in the discrete system can have a frequency comparable to the frequency of the Stokes waves itself. This suggests that the growth of these short waves may be due to resonances present in the discretised system and not present in the continuous system.

§2. The governing equations and their discretisation

The equations of motion of the interface between two incompressible inviscid fluids are equations for the evolution of the interfacial shape $z(\xi,t)$ and a interfacial vortex sheet strength $V(\xi,t)$. Here z is a complex coordinate $x + iy$ referred to fixed rectangular axes with Oy vertically upwards and ξ is a Langrangian parameter, chosen so that the point $z(\xi,t)$ moves with the average of the fluid velocities on the two sides of the interface. It will convenient to call a point of the interface with fixed ξ an "interfacial particle". It follows from this definition of ξ that if ω_+ and ω_- are the complex potentials in the upper and lower fluid

$$\frac{\partial z^*}{\partial t}(\xi,t) = \tfrac{1}{2}\left(\frac{\partial \omega_+}{\partial z} + \frac{\partial \omega_-}{\partial z}\right)_{z=z(\xi,t)} \tag{2.1}$$

But this mean value can be expressed as a Cauchy principal-value integral of the vortex sheet strength to give

$$\frac{\partial z^*}{\partial t}(\xi,t) = Q(\xi,t) \equiv \frac{1}{4\pi i} \int_0^{\xi_0} V(\xi',t)\cot[\tfrac{1}{2}(z(\xi,t)-z(\xi',t))]d\xi', \tag{2.2}$$

where a spatial periodicity 2π has been imposed on the interface corresponding to an increment ξ_0 in the Lagrangian parameter; thus for all ξ

$$\left.\begin{aligned} z(\xi+\xi_0,t) &= z(\xi,t) + 2\pi \\ V(\xi+\xi_0,t) &= V(\xi,t) \end{aligned}\right\} \tag{2.3}$$

If $V(\xi,t)$ were independent of t and thus known from the initial conditions (2.2) becomes the Birkhoff-Rott equation. The problem here is to find $V(\xi,t)$.

The choice of interfacial parameter to conform to equation (2.1) means that there is no convection of vorticity along the interface past the interfacial particle, ξ = constant, so that any change of vorticity in the element $(\xi,\xi+d\xi)$ is due to the production of vorticity by baroclinic action (Prandtl 1952, p.372). It can be shown that

$$\frac{\partial V}{\partial t}(\xi,t) = \frac{1}{\rho_+}\frac{\partial p_+}{\partial \xi} - \frac{1}{\rho_-}\frac{\partial p_-}{\partial \xi} \tag{2.4}$$

where p is pressure and ρ is density and $+$ refers to the upper and $-$ to the lower fluid. Now p_- is determined in terms of p_+ by the dynamic boundary condition at the interface and the pressure p_+ in the upper fluid can be calculated from Bernouilli's theorem. This is not completely straightforward, however, since time derivatives at constant location in space rather than at constant ξ are needed. If one defines

$$\Phi_+(\xi,t) = \phi_+(z(\xi,t),t), \tag{2.5}$$

where $\phi = \mathrm{Re}(\omega)$ is the velocity potential, then

$$\frac{\partial \Phi_+}{\partial t}(\xi,t) = \frac{\partial \phi_+}{\partial t} + \tfrac{1}{2}(\nabla\phi_+ + \nabla\phi_-)\cdot\nabla\phi_+ \quad ; \tag{2.6}$$

this equation enables the time derivative of the velocity potential at a fixed point to be evaluated. Straightforward algebra now leads to

$$\frac{\partial V}{\partial t}(\xi,t) = \left(1-\frac{\rho_+}{\rho_-}\right)\left[-\frac{\partial}{\partial t}\mathrm{Re}\left(Q_+\frac{\partial z}{\partial \xi}\right) + \frac{\partial}{\partial \xi}\left(\tfrac{1}{2}\mathrm{Re}(Q_+Q_-^*) - g\,\mathrm{Im}(z)\right)\right] + \frac{T}{\rho_-}\frac{\partial}{\partial \xi}\left(\frac{1}{r}\right), \tag{2.7}$$

where

$$Q_{\pm} = Q \mp \tfrac{1}{2} V \left(\frac{\partial z}{\partial \xi}\right)^{-1} \tag{2.8}$$

and where T is the interfacial tension and $r(\xi,t)$ the interfacial radius of the curvature, counted positive when the centre of curvature is in the upper fluid. It is worth noting that Φ itself has been eliminated by using

$$\frac{\partial^2 \Phi}{\partial t \partial \xi} = \frac{\partial}{\partial t}\left(\mathrm{Re}\left(Q_+ \frac{\partial z}{\partial \xi}\right)\right) , \tag{2.9}$$

so that the calculation of the multi-valued velocity potential is avoided.

One special case of equation (2.7) may be mentioned. If $T = 0$ and the limit $\rho_-/\rho_+ \to 1$ $g \to \infty$ and

$$(1 - \rho_+/\rho_-)g \to G \tag{2.10}$$

is taken then

$$\frac{\partial V}{\partial t}(\xi,t) = - G \frac{\partial y}{\partial \xi} \tag{2.11}$$

This is the Boussinesq limit and equations (2.2) and (2.11) were used by Hill (1974) in a study of buoyant trailing vortices.

Before equations (2.2) and (2.7) can be integrated numerically they must be discretised in both ξ and t. Suppose that initially the wave is defined by the coordinates of N arbitrarily selected interfacial particles $z_s(0)$ with strengths $V_s(0)$, where $s = 1,2,\ldots,N$ and where the particles lie along the interface in order of increasing s. These interfacial particles correspond to the point-vortices of Rosenheads (1931) method. The particle label s itself will serve as a discrete Lagrangian parameter, so that, at time t, the wave is defined by point vortex positions $z_s(t)$ and strengths $V_s(t)$. Derivatives with respect to the Lagrangian parameter are expressed by finite differences with respect to s. The Cauchy principal value integral is evaluated by a slight extension of Van der Vooren's (1980) method, so that

$$Q_s = \frac{1}{4\pi i} \sum_{\substack{r=1 \\ r\neq s}}^{N} V_r \cot\left[\tfrac{1}{2}(z_s - z_r)\right] + \frac{i}{2\pi}\left(\frac{DV_s}{Dz_s} - \tfrac{1}{2}\frac{DV_s D^2 V_s}{(Dz_s)^2}\right) \tag{2.12}$$

where the explicit t dependence is dropped for clarity and where D^p denotes a finite difference estimate of the pth derivative with respect to s. In practice, fourth-order, central, differences were used, the periodicity conditions

$$\begin{aligned} z_{s+N}(t) &= z_s(t) + 2\pi \\ V_{s+N}(t) &= V_s(t) \end{aligned} \tag{2.13}$$

being used to used to form differences at the terminal vortices, $s = 1,2$ and $s = N-1,N$.

The equations take the form

$$\frac{dz_s^*}{dt} = Q_s \tag{2.14}$$

$$\frac{dV_s}{dt} = -\left(1 - \frac{\rho_-}{\rho_+}\right)\frac{d}{dt}\left(\mathrm{Re}(Q_{+s} Dz_s)\right) + DB_s \tag{2.15}$$

where

$$B_s = (1 - \frac{\rho_-}{\rho_+})(\tfrac{1}{2}\mathrm{Re}(Q_{+s}Q_{-s}^*) - g\,\mathrm{Im}(z_s)) + \frac{T}{\rho_-}\,r_s^{-1} \qquad (2.16)$$

To discretise with respect to time equations (2.14) and (2.15) are integrated over the interval $(t, t+\delta t)$ to get estimates

$$z_s^*(t+\delta t) = z_s(t) + (\theta\,Q_s(t) + (1-\theta)Q_s(t+\delta t))\delta t \qquad (2.17)$$

and

$$V_s(t+\delta t) = V_s(t) - (1 - \frac{\rho_-}{\rho_+})(\mathrm{Re}(Q_{+s}Dz_s)_{t+\delta t} - \mathrm{Re}(Q_{+s}Dz_s)_t) +$$

$$+ (\phi\,DB_s(t) + (1-\phi)DB_s(t+\delta t))\delta t. \qquad (2.18)$$

The weightings θ and ϕ of old and new values in the integrated terms have been introduced to permit a stability analysis, which is given in the next section. However, even if old values $\theta = \phi = 1$ are used wherever possible, new values occur on the right of equation (2.18) through the term $(Q_{+s}Dz_s)_{t+\delta t}$. The problem is not one of marching and iteration must be carried out at each time step. This will be described in §4, where a practical integration scheme is defined.

§3. The stability properties of the scheme in the case of infinitesimal waves

The undisturbed interface can be described by equally spaced point vortices of zero strength, at a spacing h. When the surface is slightly disturbed

$$z_s(t) = sh + \tilde{z}_s(t) \qquad (s=1,2,\ldots,N) \qquad (3.1)$$

and

$$V_s(t) = \tilde{V}_s(t) \qquad (s=1,2,\ldots,N) \qquad (3.2)$$

where $\tilde{z}$ and $\tilde{V}$ are small quantities. The periodicity conditions (2.3) reduce to

$$\tilde{z}_{s+N}(t) = \tilde{z}_s(t)$$
$$\tilde{V}_{s+N}(t) = \tilde{V}_s(t) \qquad (3.3)$$

and

$$Nh = 2\pi, \qquad (3.4)$$

where, as in §2, there are N point vortices per wave.

The disturbance will be taken to be a single spatial Fourier component of spatial period 2π, so that

$$\tilde{z}_s(t) = \alpha(t)e^{is\psi} + \beta(t)e^{-is\psi} \qquad (3.5)$$

and

$$\tilde{V}_s(t) = \delta(t)e^{is\psi} + \delta*(t)e^{-is\psi} \qquad (3.6)$$

where

$$\psi = \frac{2\pi}{N} \qquad (N=2,3,\ldots) \qquad (3.7)$$

If (3.5) and (3.6) are substituted into the discretised equations (2.17) and (2.18) and nonlinear terms rejected it can be shown that, provided the initial displacements $\tilde{z}_s(0)$ are vertical,

$$\alpha(t) + \beta^*(t) = 0,\tag{3.8}$$

so that the displacements remain vertical, while

$$\alpha(t) = A\mu_1^n + B\mu_2^n,\tag{3.9}$$

where A and B are constants, $t = n\delta t$ and μ_1 and μ_2 are the roots of the quadratic equation

$$(\mu-1)^2 = -\eta^2\delta t^2(\theta+(1-\theta)\mu)(\phi+(1-\phi)\mu);\tag{3.10}$$

$\delta(t)$ is given similarly. The constant η is defined by

$$\eta^2 = \frac{g}{h}(1+\frac{\rho_+}{\rho_-})^{-1}\{(1-\frac{\rho_+}{\rho_-}) + \frac{2T}{\rho_-gh^2}f_1(\psi)\}\,f_2(\psi)\tag{3.11}$$

where the functions f_1 and f_2, which depend on the finite difference estimates of D and D^2, are, for fourth-order centred differences

$$f_1(\psi) = \frac{5}{4} - \frac{4}{3}\cos\psi + \frac{1}{12}\cos2\psi\tag{3.12}$$

$$f_2(\psi) = (1-\frac{\psi}{\pi}+\frac{4}{3\pi}\sin\psi - \frac{1}{6\pi}\sin2\psi)(\frac{4}{3}\sin\psi-\frac{1}{6}\sin2\psi)\tag{3.13}$$

The analysis leading to these results is lengthy, but it is a straightforward extension of Von Karman's discussion of the stability of a uniform linear array of point vortices (Lamb 1932, p.225). Thus the details will not be given.

It can be verified that as $\delta t \to 0$ and $N \to \infty$, these results reduce to those for the continuous case. If $\eta\,\delta t \ll 1$, then it can be shown from (3.10) that

$$\mu^n = e^{\pm i\eta t}(1+\tfrac{1}{2}(\theta+\phi-1)\eta^2t\delta t + O(\eta^3t\delta t^2)).\tag{3.14}$$

Expansion of $f_1(\psi)$ and $f_2(\psi)$ shows that

$$\eta^2 = g(\frac{\rho_--\rho_+}{\rho_-+\rho_+}) + \frac{T}{\rho_-+\rho_+} + O(N^{-4}),\tag{3.15}$$

so that as $\delta t \to 0$ and $N \to \infty$ equation (3.4) reduces to the result for a wave of unit wavenumber (Lamb 1932, p.459). The spatial truncation error is small because of the use of fourth-order centred differences.

It is helpful to discuss the stability of the numerical scheme with the aid of the (θ,ϕ) plane. If

$$\eta^2\delta t^2 < 4/(\theta-\phi)^2 \qquad\qquad 0 \leqslant \phi\quad\theta \leqslant 1\tag{3.16}$$

then the quadratic equation (3.10) has complex conjugate roots. The values of (θ,ϕ) of interest lie in the square $0\leqslant \phi,\theta\leqslant 1$ and if, as will be assumed henceforth, $\eta^2\delta t^2 < 4$ the inequality (3.17) is satisfied throughout the square. It is easily shown that, in this case,

$$|\mu|^2 = (1+\eta^2\delta t^2\theta\phi)/(1+\eta^2\delta t^2(1-\theta)(1-\phi))\tag{3.17}$$

so that

$$|\mu| = 1 \text{ on } \theta+\phi = 1\tag{3.18}$$

and

$$|\mu| < 1 \quad \text{for} \quad \theta + \phi < 1 \tag{3.19}$$

A sufficient condition for stability is $|\mu| \leqslant 1$ and this is satisfied on and below a diagonal of the basic square. The scheme is stable and non-dissipative <u>on</u> the diagonal and the temporally centred scheme $\theta = \phi = \frac{1}{2}$ has least temporal truncation error, in view of equation (3.14). This choice was made in practice, with results described in §4.

§4. Finite amplitude instability

A program was written to integrate (2.17) and (2.18) was written, the fully-time centred choice $\theta = \phi = \frac{1}{2}$ being made for the reasons described in §3. The right-hand side of each equation involves in a nonlinear fashion $z_s(t+\delta t)$ and $V_s(t+\delta t)$ and this difficulty was dealt with by functional iteration. An estimate $\bar{z}_s^{(0)}(t+\delta t)$ was computed from previously calculated values by

$$\bar{z}_s^{(0)}(t+\delta t) = 4z_s(t) - 6z_s(t-\delta t) + 4z_s(t-2\delta t) - z_s(t-3\delta t) . \tag{4.1}$$

This estimate and a similar one $\bar{V}_s^{(0)}(t+\delta t)$ for the strengths were used to calculate the left-hand side of (2.18) and (2.19). These – hopefully improved – estimates $\bar{z}_s^{(1)}(t+\delta t)$ and $\bar{V}_s^{(1)}(t+\delta t)$ were substituted in the right-hand side of (2.18) and (2.19) and the process continued until convergence was achieved. Convergence was taken to be complete when

$$\frac{1}{N} \sum_{s=1}^{N} (|\bar{z}_s^{(p+1)} - \bar{z}_s^{(p)}|/H_m + |V_s^{(p+1)} - V_s^{(p)}|/V_m) < .1\delta t^3 \tag{4.2}$$

where $H_m = \max|y_r - y_s|$ and $V_m = \max|V_s|$; there is no point in iterating past this point since the local truncation error in (2.18) and (2.19) is $O(\delta t^3)$ when $\theta = \phi = \frac{1}{2}$. No difficulties with convergence were experienced, one or two iterations only being required in most cases.

The amplitude expansion for the Stokes wave given by Schwarz (1974) was used to test the program. Schwarz gives the wave shape and propagation speed of the wave as a power series in the semi-amplitude a of the wave, where

$$a = 0.5 \text{ (height of crest – height of trough)}; \tag{4.3}$$

Thus recalling that the wavelength has been taken to be 2π, $a = .4436$ is the Stokes wave of greatest height in these units. Schwarz does not give the vortex sheet strength and this was computed numerically.

Schwarz's results were used to compute initial data for the program and subsequently the calculated wave was compared with the analytical solution, the quantity

$$\varepsilon = \frac{1}{2Na} \sum_{s=1}^{N} |y_s(\text{calc}) - y_s(\text{theory})| \tag{4.4}$$

being a convenient measure of the numerical performance. In all calculations $\rho_+ = 0$, $\rho_- = 1$, $g = 1$ and $T = 0$, so that the wave speed c is given by

$$c^2 = 1 + a^2 + \ldots \quad ; \tag{4.5}$$

thus $c \approx 1$. With $\delta t = .025$, $N = 18$ and $a = .2$ it was found that at $t = 6.0$, when the wave had moved roughly one wavelength, $\varepsilon = .004$, a satisfactory

result bearing in mind the modest number of point vortices used. However
at t = 12.0 ε = .025 and the calculation has failed. Plotting the results
revealed the appearance of a zig-zag pattern in the point vortices.

The appearance of a zig-zag pattern of Lagrangian points is a
characteristic mode of failure of Lagrangian methods and Longuet-Higgins
and Cokelet (l.c.) encountered it. This suggests that spurious modes of
high wavenumber are entering the Fourier decomposition of the results as has
been previously discovered by Fornberg (private communication).

To study this further a fast Fourier transform of the results was taken
as the calculation proceeded. A typical result is shown in Figure 1, for
δt = .1 N = 64 and a = .018 at time t = 140. The Fourier coefficient
Z_n is defined by

$$Z_n = \sum_{s=1}^{N} y_s \, \exp(\frac{2\pi i n s}{N}) \quad (n=1,2,\ldots,N) \tag{4.6}$$

The value of n = N/2 corresponds to a zig-zag of constant amplitude and it
may be noted that this is not the largest component, which is n = 29 in this
case. Some components grew exponentially with time and some grew in an
oscillatory fashion, but with the same average rate, as estimated by a
least squares fit. The growth of the mode n = N/2 - 1 is shown in
Figure 2, for N = 32, 64 and 128. The growth rate which is much larger than
that of the physical short wave instabilities is roughly linear in the wave
semi-amplitude a showing that the instability is nonlinear and increases
rapidly with N. It was not significantly changed by changing δt, suggesting
that the growth of these spurious modes is not a consequence of the temporal
discretisation. However, a change in δt affected the detailed results,
possibly because δt affects the initial excitation.

A significant feature is the appearance of a threshold amplitude below
which the mode did not grow. The case N = 64 and a = .005 and N = 32 and
a = .005 exhibited practically no growth in the time interval $0 \leqslant t \leqslant 368$.
Unfortunately, the expense of such long runs prevented the value of the
threshold amplitude being determined. No threshold was observed for N = 128,
but only a few such runs were carried out because of their cost.

A possible explanation of these reuslts is the introduction of
resonances by the distortion by the discretisation of the dispersion curve.
The normal mode in which

$$y_s = \exp(\frac{2\pi i n s}{N}) \qquad (n=0,1,\ldots,N/2) \tag{4.7}$$

has a frequency η_n given by

$$\eta_n^2 = \frac{N}{2\pi} \, f_2(\frac{2\pi n}{N}) \qquad (n=0,1,\ldots,N/2) \tag{4.8}$$

where f_2 is defined in equation (3.13); this is easily deduced from (3.11).
A comparison of η_n with the dispersion curve for deep water waves is given
in Figure 3. Note that $\eta_{N/2} = 0$ so that the pure zig-zag wave has zero
frequency, a point noticed by Baker (private communication). Clearly the
discretisation by distorting the dispersion curve has introduced a
possibility of resonance between the dominant n = 1 component of Stokes
wave and modes with n $\sim$ N/2. If N is large and N/2-n is finite (4.8) can
be approximated by

$$\eta_n = \frac{4\sqrt{5}}{3\sqrt{N}} \quad (N/2-n). \tag{4.9}$$

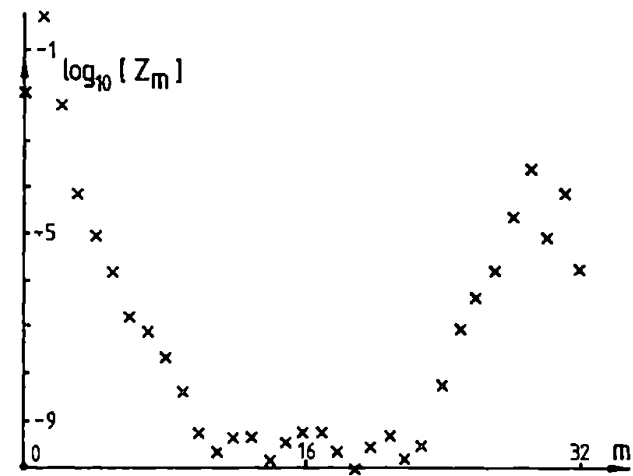

Figure 1:   Typical spectrum

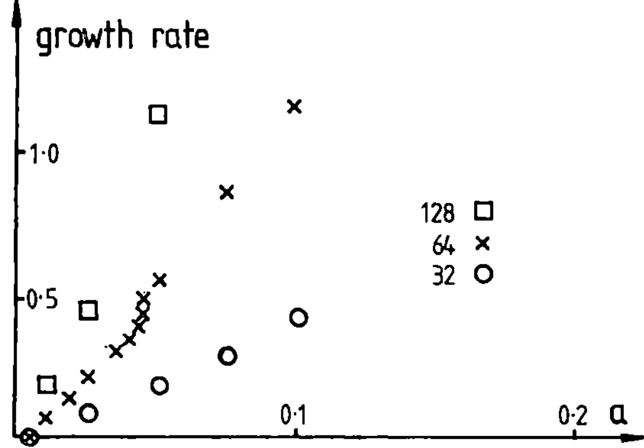

Figure 2:   Growth of mode $m = N/2 - 1$

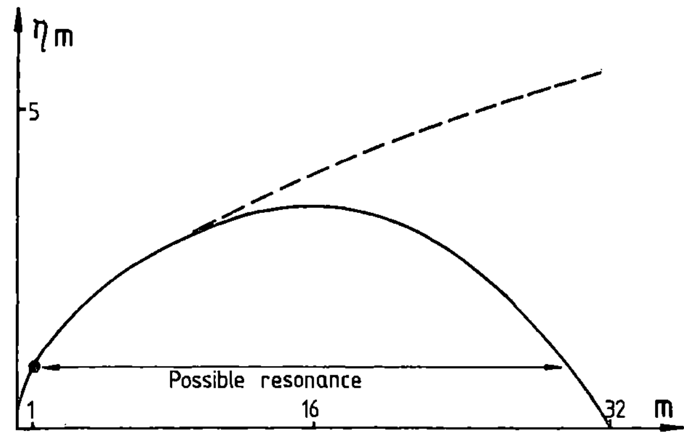

Figure 3:   Dispersion curve (solid line)

The triad of modes 1, n, n+1 can resonate if

$$\eta_1 = \eta_n + \eta_{n+1} \tag{4.10}$$

leading to

$$N/2 - n = \tfrac{1}{2} + \frac{3\sqrt{N}}{8\sqrt{5}} \tag{4.11}$$

Exact resonance cannot in general be achieved, consistent with the observation of a threshold amplitude. However the detuning is not small and the value of n, while being less than N/2 as observed, is too large in general. In any case, an analytical solution of a suitably approximated form of the nonlinear discrete equations (2.18) and (2.19) would be needed to establish the resonance explanation, which must remain speculative.

Acknowledgements

This work was carried out at the California Institute of Technology with support by the U.S. Office of Naval Research.   I thank Professor B. Fornberg and Professor P.G. Saffman for many helpful discussions and Dr J.C. Schatzman for valuable programming advice.   I am grateful to Dr H.P. Peregrine for giving me access to some of his numerical results on stationary waves.

References

Baker, G.R., Meiron, D.I. and Orszag, S.A. (1980) Phys. Fluids, 33, 1485.
Fornberg, B. (1980) SIAM, J. Sci. and Stat. Comp. 1, 386.
Hill, F.M. (1974) JFM 71, 1.
Kulyaev, R.L. (1976) J. Appl. Mech. and Tech. Phys. 16, 76.
Lamb, H. (1932) Hydrodynamics. C.U.P.
Longuet-Higgins, M.S. and Cokelet, E.L. (1976), Proc. Roy. Soc. A350, 1.
Prandtl, L. (1952), Fluid Dynamics, London, Blackie.
Pullin, D., (1982), to appear.
Rosenhead, L. (1931) Proc. Roy.Soc. A134, 170.
Schwarz, L.W. (1974), JFM, 62, 533.
Van der Vooren, A.I. (1980), Proc. Roy. Soc. A373, 67.
Zarodny, S.J. and Greenberg, M.D. (1973), J. Comp. Physics 11, 440.

VORTICES IN TURBULENCE

by

A.E. Perry, M.S. Chong and T.T. Lim[†]
Reader and Research Fellows respectively
Department of Mechanical Engineering
University of Melbourne
Parkville 3052, Victoria, Australia

INTRODUCTION

Some recent work of Perry and Chong (1981, 1982) is reviewed concerning
the structure of wall turbulence. In this work it was assumed that wall
shear flow can be modelled as a "forest" of hair-pin or "Λ" shaped vortices
which are embedded in an irrotational flow field. By making certain plausible
assumptions regarding the velocity and length scale distributions of these
vortices, the mean flow field, broad-band turbulence intensity distributions
and spectra are derived which are consistent with measured data. Much of this
modelling was inspired by the smoke tunnel observations of Head and
Bandyopadhyay (1981), Perry, Lim and Teh (1981) and Perry and Chong (1982).
Using similar concepts and experimental techniques, this work has been
extended to plane mixing layers. This new work is reported here and
comparisons are made with wall turbulence. By the use of spectra, it is
shown that significant differences exist between the two types of turbulence.
Possible reasons for these differences are explored.

1. WALL TURBULENCE

One of the most significant findings in the work of Head and Bandyopadhyay (1981) was that wall turbulence as shown by smoke studies has a
"granular" structure with a characteristic direction. A sheet of laser light
normal to the wall and aligned with the flow direction shows the smoke
filled layers to have an array of structures which lean approximately 45°
in the downstream direction as illustrated schematically in figure 1(a).
These structures remain approximately straight and are convected with a uniform velocity in the streamwise direction. Two further sheets of laser light
were used in the cross-stream direction and these are labelled 1 and 11 in
figure 1(a). Sheet 1 was aligned approximately with the structures and sheet
11 was normal to the structures. The sectioning as viewed looking upstream
are shown in figure 1(b) and (c). Loops in figure 1(b) and mushroom-like
patterns in figure 1(c) are consistent with the hair-pin vortices. However,
the most convincing evidence for the existence of a "granular" structure
with a characteristic direction is the apparent downward movement of the
mushrooms in figure 1(c) as the structures in figure 1(a) move from left to
right and cut the laser sheet 11. The cine film of Bandyopadhyay and Head (1979)
shows this downward movement very clearly. This same flow visualization
technique was used by the authors in the plane mixing layer and is described
later.
Perry and Chong (1982) have shown by a Biot-Savart law calculation that
an isolated Λ vortex interacts with its image in such a way that it stretches
at a uniform rate and the legs of the vortex are brought together. This is

† Present address: NASA Ames Research Centre, Moffett Field, CA 94035

illustrated in figure 2(a). It has been shown by Townsend (1951 a,b),
Batchelor (1967) and Perry and Chong (1982) that the diameter of a vortex
rod remains constant only if a marked length of rod is increasing
exponentially with time. Perry and Chong (1982) have shown that a uniform
stretching means that viscous diffusion will ultimately dominate over the
stretching effect and the rod diameter will grow. This, coupled with the
legs of the Λ vortex being brought together leads to the idea of vorticity
cancellation and ultimate "death" of the eddy. Thus in wall turbulence in
a pipe where the vortex population is stable, there must be a continual
"birth" of new eddies from the viscous sublayer material and this birth
rate must balance the death rate. The debris of dead eddy material is
assumed to be irrotational.

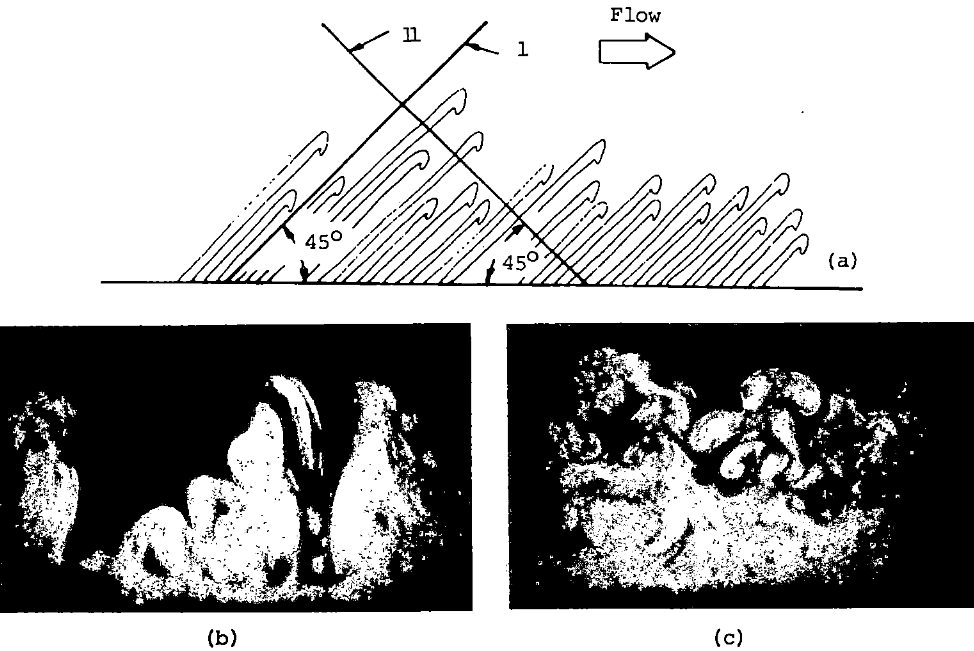

Figure 1.   Smoke observations of Head and Bandyopadhyay (1981)
Turbulent boundary layer R_θ = 600.
(a) Schematic section aligned with flow and normal to wall showing
cross-stream laser sheets.
(b) Laser sheet 1 section.   (c) Laser sheet ll section.

Townsend (1976) and Perry and Chong (1982) have shown that if one is
to obtain a logarithmic mean velocity distribution and a corresponding
region of constant Reynolds shear stress, it is necessary to have a hierarchy
of scales of geometrically similar vortices all with a constant characteristic
velocity scale but the length scale must be distributed with a "logarithmic-
like" probability.

One model considered by Perry and Chong was the so-called "quantum
jump" model where the length scales have a discrete p.d.f. and go in a
geometrical progression with a factor of 2. This is shown schematically
in figure 2(b). Of course, "jitter" will tend to smear out this discrete
distribution to give a continuous inverse power law p.d.f. (or logarithmic
probability). The first hierarchy scales with the Kline scaling (in figure
2(b) $\lambda = 100\nu/U_\tau$, where U_τ is the friction velocity and ν is the kinematic

viscosity,(see Kline (1967)). How the hierarchies are formed is still a
mystery. Vortex pairing is one possible mechanism and is consistent with
the quantum jump model and the necessary scaling. If the scale doubles, the
circulation of the vortex doubles, giving a constant characteristic velocity
for each hierarchy. This velocity scale is defined later.

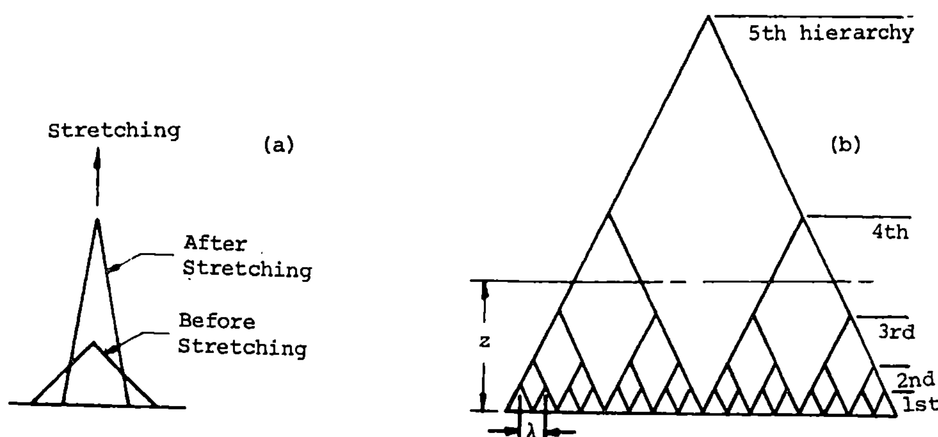

Figure 2.   (a)   Vortex stretching.   (b) Schematic representation of
discrete system of hierarchies.   Both diagrams are
views looking upstream.

Perry and Chong found that spectra could be generated using potential
flow vortices by the use of eddy "signatures" obtained from the Biot-Savart
law. Figure 3 illustrates a typical cross-stream row of "representative"
eddies for the first and second hierarchies. For the purpose of computing
velocity signatures and hence spectra, the Λ vortices are treated as parallel
vortex pairs of infinite extent and the cross-stream component of vorticity
is ignored. Actually, in the analysis of Perry and Chong (1981, 1982) the
rods were allowed to slant in a more general way but for simplicity, the
above arrangement is used here. It seems to be more appropriate for the plane
mixing layer. This simplification does not affect the functional forms of the
resulting spectra. In wall turbulence, the streamwise spacing of the various
eddy rows increases in a geometrical progression with a factor of 2. In the
first hierarchy this streamwise spacing scales with y_0.

Perry and Chong (1981, 1982) found by an approximate analysis (using the
simple discrete model) that the power spectral density for the streamwise
velocity fluctuations u_1 is given by

$$\frac{\Phi(k_1 y_0)}{v_0^2} \sim \sum_{n=1}^{N_0} \frac{1}{k_1 y_0} \left\{ \exp(-2k_1 y_0 2^{n-1}) - \exp(-2k_1 y_0 \frac{y_1}{y_0} 2^{n-1}) \right\} \qquad (1)$$

where $\Phi(k_1 y_0)$ is the energy per unit nondimensional wavenumber $(k_1 y_0)$, k_1
could be   regarded as the streamwise wavenumber, y_0 is the smallest scale of
the smallest observed hierarchy and N_0 is the number observed hierarchies.
The velocity scale v_0 is constant for all hierarchies and $v_0 \sim K/y_0$ where K
is the circulation of vortices in the smallest observed hierarchy.# For
wall turbulence $v_0 \sim U_\tau$, the friction velocity. Note that equation (1) is

---

\#   Throughout this paper the symbol $\sim$ means "scales with" or "is porportional
to with a universal constant involved".

108

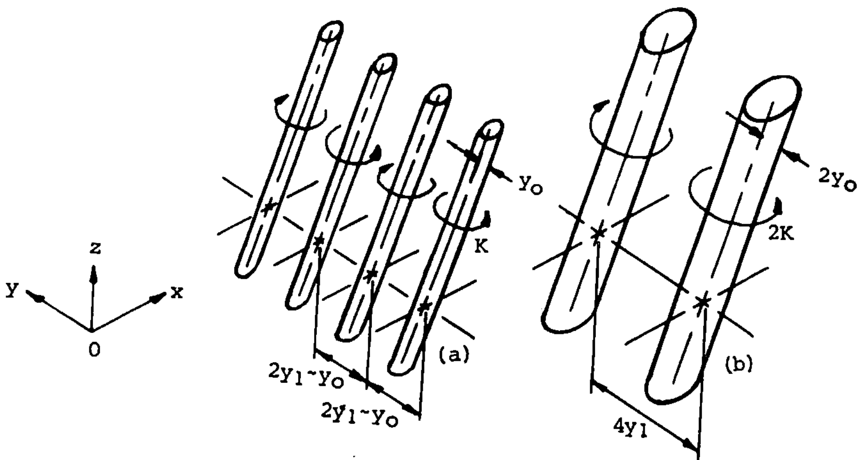

Figure 3. Representative eddies for the first two hierarchies.
(a) First hierarchy.   (b) Second hierarchy.

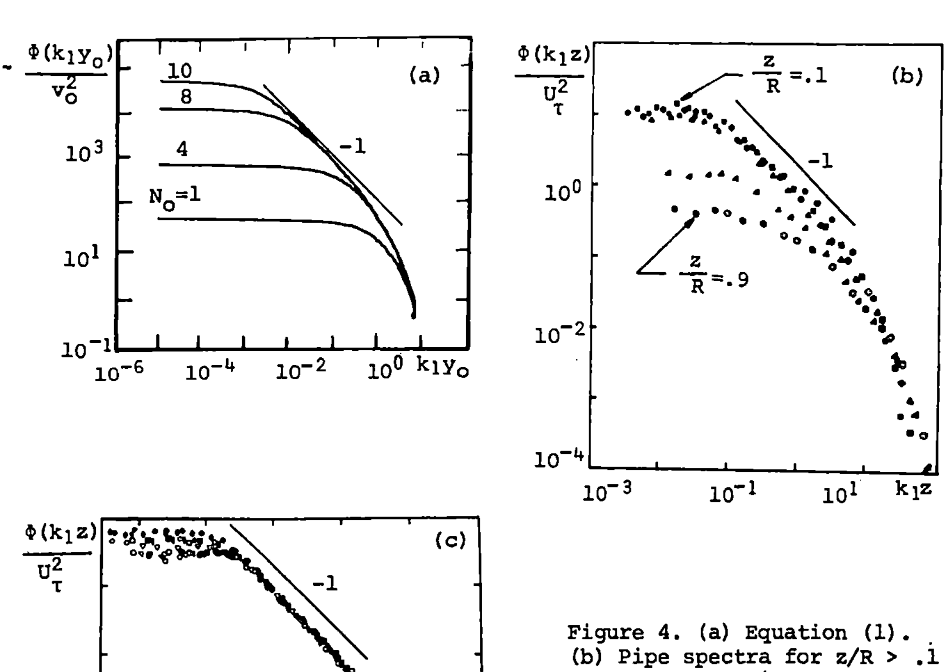

Figure 4. (a) Equation (1).
(b) Pipe spectra for z/R > .1
(c) z/R < .1; z^+ > 100.

R is pipe radius and $z^+ = \dfrac{zU_\tau}{\nu}$.

Perry and Abell (1974,1977)
In 4(a), length of -1 region
increases with N_o.

general for all distances z from the wall and in wall turbulence y_0 is not
necessarily the scale of the first hierarchy, but is the scale of the smallest
observed hierarchy. For instance, for the value of z shown in figure 2(b), the
smallest observed hierarchy is the fourth hierarchy and N_0, the number of
observed hierarchies is 2. The scale y_0 (with jitter) is proportional to z. Note
also that the number N_0 of observed hierarchies diminishes with z. Equation (1)
is shown plotted in figure 4(a) for various N_0 where y_1/y_0 has been arbitrarily
taken to be 10 and is constant for all the hierarchies. Also shown in figure
4(b) and (c) are some experimental results of Perry and Abell (1975,1977) for
pipe flow turbulence. It can be seen from figure 4(b) that the number of
observed hierarchies diminishes with increasing z from the wall. Figure 4(c)
shows data for the logarithmic wall region and the appropriate spectral law
for the energy containing region is a -1 power law. The scaling used in
figures 4(b) and (c) for collapsing the data is completely consistent with
equation (1). The theoretical spectra appear to have all the correct features.

2.   THE PLANE MIXING LAYER
     A plane mixing layer was chosen as an example of turbulence which possesses
certain simple properties. One property can be derived from the Reynolds
number similarity hypothesis of Townsend (1976). That is, if viscosity does
not enter the problem explicitly, then the growth rate is linear with x
(see figure 5) and the flow is self preserving. This means that mean relative
motions and the energy containing turbulence quantities will collapse when
plotted and non-dimensionalized with one characteristic velocity scale U_0
(the velocity jump across the layer) and a local length scale l_s which grows
linearly with x. Figure 5 shows the geometry and definition of the various
terms. Also shown are the mean flow and broad-band turbulence intensity
results. Self preserving flow appears to be applicable.
     Figure 6(a) shows the flow with smoke injected when viewed side on (the
splitter plate is horizontal). Large scale roll-ups can be seen to be vaguely
similar to those observed by Brown and Roshko (1974) but are not so well defined
(i.e. they appear to be more "turbulent" even though the boundary layers on the
splitter plate were laminar). Figure 6(b) shows a plan view and as one goes
downstream the structures grow and lose their spanwise uniformity. The
structures appear to have a granular appearance with a characteristic
direction implying that there exists longitudinal vortices "riding" on the
cross-stream roll-ups. This is consistent with the observations of
Breidenthal (1981) and recent observations of Roshko (1981)[*]. Figure 6(c)
shows a cross section of smoke given by laser sheet 1 (see figure 5(a)) and
figure 6(d) gives the cross-section seen with sheet 11. Although the mushrooms
and loops do not show up as clearly as in the Head and Bandyopadhyay (1981)
photographs for boundary layers, a cine film of laser cross-sections 11
shows a definite downward movement of the patterns. Thus it seems plausible
that a large amount of turbulent kinetic energy is given by longitudinal
vortices with a characteristic direction. Longitudinal vortex loops can be
seen in figure 6(b). It will be assumed that most of the energy comes from
longitudinal vortex pairs.
     Figure 7(a) shows experimental spectra taken on the plane z = 0.
This is defined to correspond with $\Delta U/U_0 = 1/2$. Here $\Phi(k_1)$ is the energy per
unit wavenumber k_1 ($=2\pi f/U_c$) where f is the frequency in Hz. We will confine
most of our attention to $z = 0$ and assume that the structures are convected at
$U_c = (U_1 + U_2)/2$.

*   Private communication

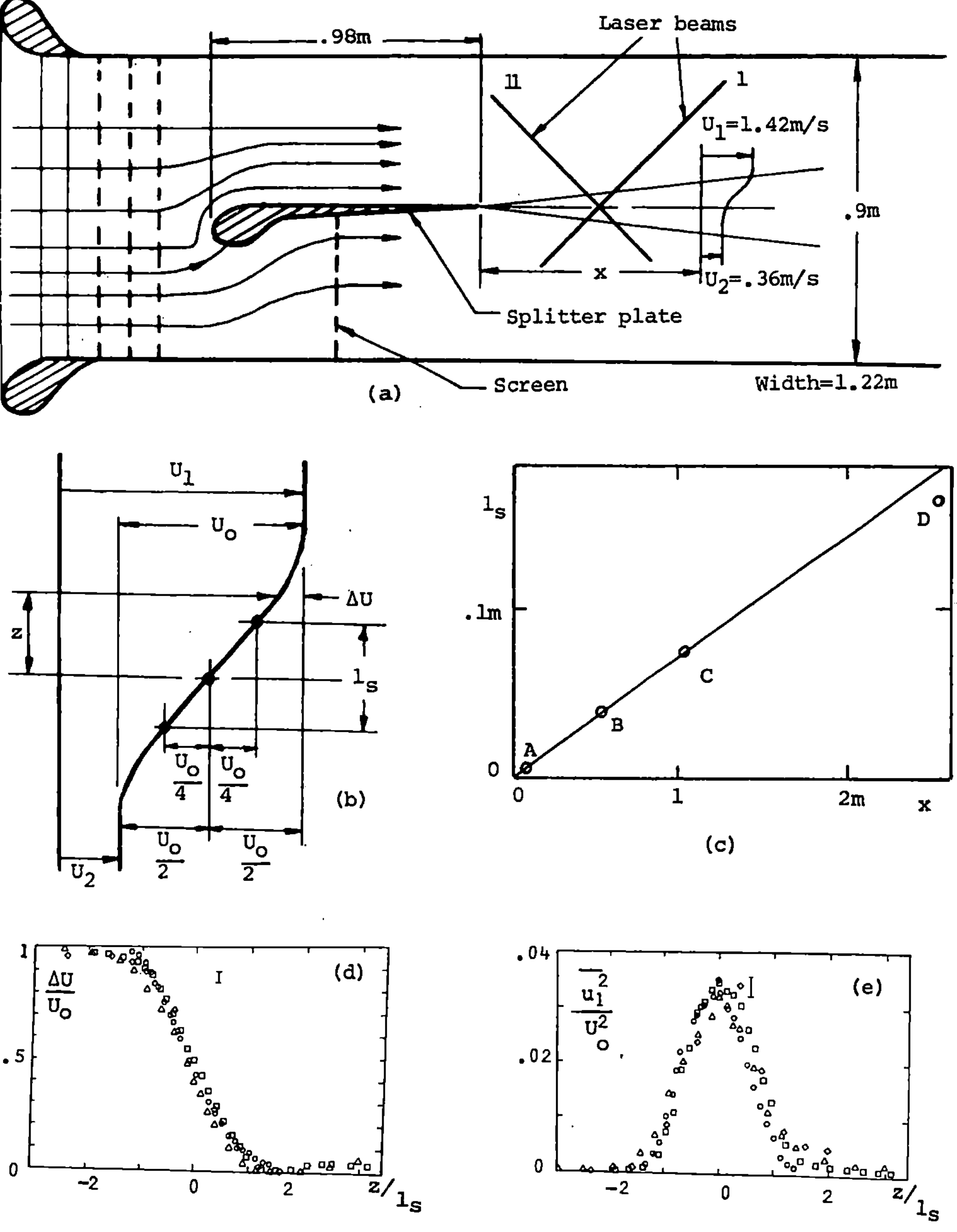

Figure 5. Plane mixing layer.
  (a) Experimental arrangement.
  (b) Definition of terms with mean velocity profile.
  (c) Growth rate; A=.06m; B=.52m; C=1.0m; D=2.5m are x values.

$$R_1 = \frac{xU_O}{\nu} = 0.70 \times 10^5 /m \qquad R_2 = \frac{xU_c}{\nu} = 0.58 \times 10^5 /m$$

  (d) Mean velocity profiles (e) Turbulence energy profiles.
      Single normal hot-wire was used calibrated statically.

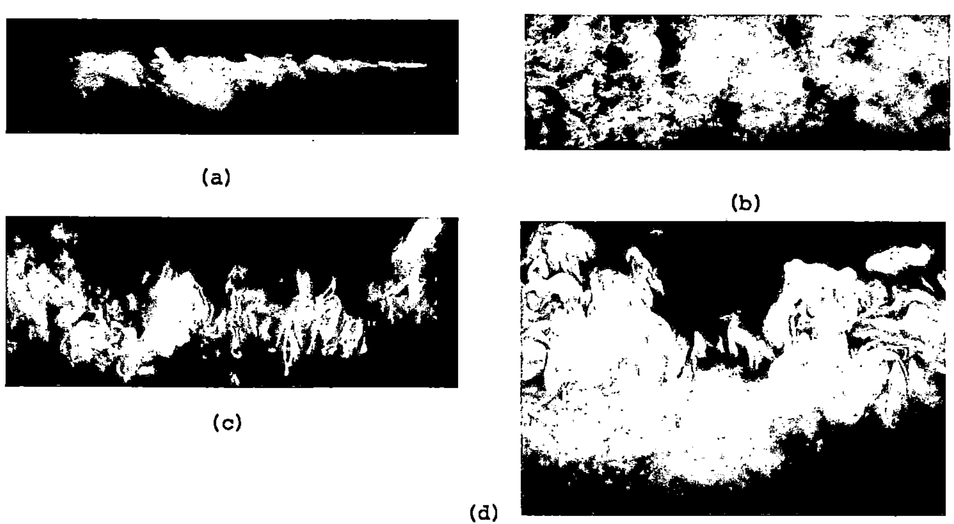

Figure 6. Smoke observations of plane mixing layer.
(a) Section normal to horizontal splitter plate and aligned
with the flow. Flow from right to left.   (b) View looking
down from above. Flow from left to right.   (c) Laser section 1.
(d) Laser section II .   Distance from splitter plate=.75m.

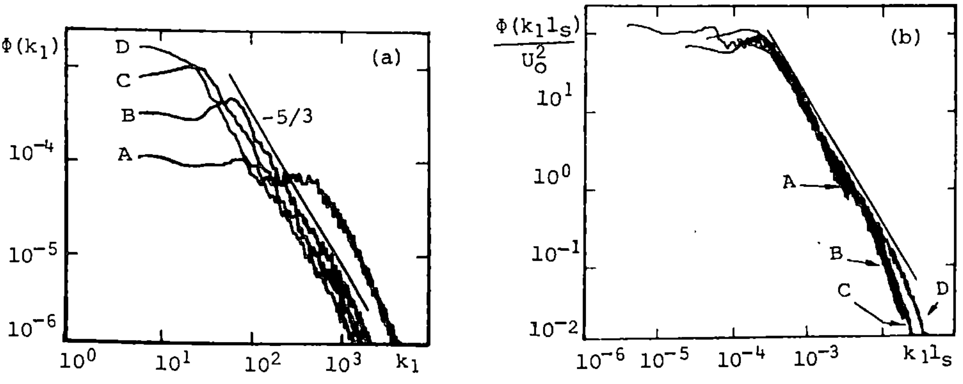

Figure 7.   Experimental spectra.   (a) Normalized spectra (see text).
(b)   Same spectra nondimensionalized with U_o and l_s.

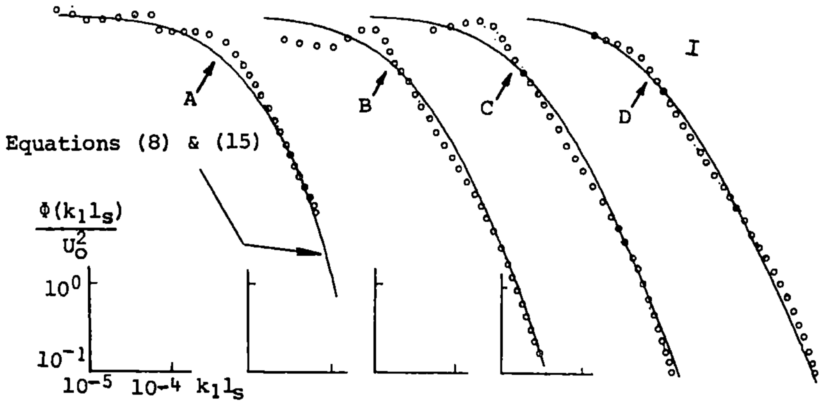

Figure 8.   Experimental spectra compared with theory. Data same
as in figure 7. Theoretical spectra are for the
same cases given in figures 9(a) to 9(d). $y_1/y_o = 10$
This parameter has only a weak effect on spectra shape.

It can be seen that for much of the spectrum, the appropriate power
law is close to -5/3. The number of hierarchies (i.e. the length of the
-5/3 law) increases with streamwise distance. $\Phi(k_1)$ is defined such that

$$\int_o^\infty \Phi(k_1)\, dk_1 = \overline{u_1^2}$$

where the overscore denotes a time average.

Figure 7(b) shows spectra for 4 values of x non-dimensionalized by the
characteristic length l_s of the shear layer and the characteristic velocity
U_o. Here $\Phi(k_1 l_s)$ is the energy per unit non-dimensional wavenumber $k_1 l_s$.
Here a reasonable collapse can be seen to occur in the energy containing
wavenumber region including the -5/3 region. This is consistent with the
self preserving flow hypothesis. Figure 8 shows the plots separately and
low frequency peaks are seen probably due to periodicity of the cross-stream
vortices. These disappear as we move downstream. It will be shown later
that it is quite probable that these spectra will also collapse at the very
high and "-5/3 law" range of wavenumbers if we use the classical Kolmogorov
velocity and length scales. This collapse should occur even though it is
quite obvious from the smoke observations that local isotropy is unlikely
to be applicable. Furthermore, the Reynolds numbers were moderate.

One can see that with appropriate scaling, spectra in wall turbulence
collapse on a -1 law region and "peel off" at different wavenumbers
depending on the flow conditions and position in the flow. The corresponding
law for the plane mixing layer is close to a -5/3 law.

Possible reasons for this difference in behaviour should be pursued.
It may be many years before questions regarding this will be resolved in a
satisfactory way, if ever. However, the authors wish to outline a somewhat
speculative model which may have some element of truth regarding the above

phenomenon.

3. MIGRATION OF EDDIES IN HIERARCHY SPACE

Let it be assumed that vortex pairing is the process by which the
hierarchies are formed in both wall turbulence and in the plane mixing layer.
There is very little evidence that this process occurs in wall turbulence
but vortex pairing has been definitely observed in plane mixing layers, at
least for the spanwise vortices (e.g. see Winant and Browand (1974) and
Brown and Roshko (1974)). Recently, Perry and Tan[*]have devised an experiment
in negatively buoyant coflowing wakes where three-dimensional vortex loops
are seen to pair and the pairing process is highly controllable.

For simplicity, assume that cross-stream rows of longitudinal eddies pair.
The smallest eddies which form at the source (i.e. at the wall in the case of
wall turbulence) will be referred to as "basic eddy rows". These are the eddies
in the first hierarchy. In the case of the plane mixing layer, the spanwise
rows of longitudinal eddies which form just after the splitter plate will be
the basic eddy rows. Like in wall turbulence, we will assume that the basic
eddy rows consist of longitudinal vortex pairs and two rows somehow pair to
give a geometrically similar row of longitudinal pairs but the lateral scale
of the vortex pairs is doubled and so also is the circulation. Thus the
second hierarchy eddy row consists of two basic eddy rows. This will in
turn pair with another second hierarchy row to give a third hierarchy row
which consists of four basic eddy rows and so on. We imagine that we have
a short sampling length situated near the trailing edge of the splitter
plate and we count the number of basic eddy rows per unit length. This is
repeated many times and an ensemble average is obtained. We then move our
sampling length downstream and again count the number of basic eddy rows.
From equation (1), when it is applied to wall turbulence, it will be found
that the number of basic eddy rows per hierarchy per unit length is constant
and as a consequence of this, the contribution to the turbulence kinetic
energy is constant from each observed hierarchy. Thus the number of basic
eddy rows per unit length determines the turbulent kinetic energy. In the
plane mixing layer $\overline{u_1^2}$ is constant along the plane z = 0. Hence, the number of
basic eddy rows per unit length in our sampling length must remain invariant.
Thus no basic eddy rows die. How the flow achieves this is open to debate.
It would imply that a stretching process is at work which maintains the rods
under a continual exponential stretching. This might come from some form of
multiple folding process or perhaps from "saddle-like" critical points set
up between spanwise roll-ups (see Perry, Lim and Chong (1981) and Perry and
Watmuff (1981)). In the case of pipe turbulence, new basic eddy rows are
forming at the boundary continuously and it is necessary for eddy rows to die
so that a stable population of eddy rows exist. It can be shown from the
work of Perry and Chong (1982), that the life time of an eddy in each
hierarchy is such that half the eddy rows pair and half die.

All vorticity in the shear layer comes from the splitter plate and no
new vorticity is generated downstream since the flow is incompressible and
of constant density (see Lighthill (1963)). The flux of mean vorticity is
constant so there is no vorticity cancellation at least for the crossflow
component. As we now move downstream our sampling length will show a constant
number of basic eddy rows per unit length but the rows will have "migrated"
in hierarchy space to give a distribution of hierarchies.

Let N_n be the number of basic eddy rows per unit length in the nth
hierarchy and let M_n be the number of nth hierarchy rows. Thus

$$N_n = 2^{n-1} M_n \qquad (2)$$

* Unpublished

Let H_n be the flow rate of nth hierarchy eddy rows per unit length flowing out of the nth hierarchy into the $(n+1)$th hierarchy. If Q_n is the corresponding flow rate of basic eddy rows, then

$$Q_n = 2^{n-1} H_n \tag{3}$$

Now v_o is the characteristic velocity of the vortex pairs. This will be constant for all hierarchies and will scale with U_o

Now let it be assumed[†] that $H_n = f_1(M_n, v_o) = f_2(M_n, U_o)$ $\qquad$ (4)

and by dimensional analysis

$$H_n = \kappa U_o M_n^2 \tag{5}$$

where κ is a universal constant (assuming the behaviour of all hierarchies is universal). From equations (2) to (5),

$$Q_n = \kappa U_o N_n^2 / 2^{n-1} \tag{6}$$

Given that there is no basic eddy row death then a balance of basic eddy flow rates in a given hierarchy is given by

$$\frac{\partial N_n}{\partial t} = Q_{n-1} - Q_n \tag{7}$$

We could imagine that our sampling length is being convected downstream at a velocity U_c and that $t = x/U_c$, i.e. time t is "space-like". From equation (3) and (7)

$$\frac{\partial N_n}{\partial t^*} = \frac{1}{2^{n-1}} \left\{ 2N_{n-1}^2 - N_n^2 \right\} \tag{8}$$

where $t^* = \kappa U_o t$ and from the conservation of total number of basic eddy rows per unit length

$$N_T = \sum_{n=1}^{N_o} N_n = \text{constant} \tag{9}$$

where N_T is the total number of basic eddy rows per unit length and N_o is the total number of hierarchies. Note that on the plane $z = 0$ to which we are confining our attention, all hierarchies in the flow are being observed.

From equation (8) we have a set of simultaneous first order non-linear differential equations. For $n = 1$

$$\frac{dN_1}{dt^*} = -N_1^2 \tag{10}$$

or $\qquad\qquad N_1 = N_T / (1 + N_T t^*) \tag{11}$

where N_T, the total number of basic eddy rows $= N_1$ at $t^* = 0$.

† see section 4.

The other equations $(n > 1)$ are solved with $N_1 (t^*)$ being the forcing function for the equation $n = 2$ and with $N_2 (t^*)$ being the forcing function for equation $n = 3$ and so on.

Figure 9(a) illustrates how the basic eddy rows migrate among the hierarchies according to the above equations. From the figure it can be seen that the population of the first hierarchy is being continually depleted at the expense of the higher hierarchies in this migration game. The power spectral density from equation (1) needs to be modified to read

$$\frac{\Phi(k_1 y_o)}{U_o^2} \sim \sum_{n=1}^{N_o} \frac{f(n)}{k_1 y_o} \left\{ \exp(-2k_1 y_o 2^{n-1}) - \exp\left(-2k_1 y_o \frac{y_1}{y_o} 2^{n-1}\right) \right\} \tag{12}$$

where $f(n)$ is a weighting factor given by

$$f(n) = N_n/N_T \tag{13}$$

The spectra given by equations (12) and (13) are shown in figure 9(b) for $y_1/y_o = 10$ and are compared with the classical $-5/3$ law. Actually, it can be shown analytically that the asymptotic law for $t^* \to \infty$ for equations (12) and (13) is a -2 law (which corresponds to $f(n) \sim 2^{(n-1)}$). This differs slightly from the $-5/3/$ law.

The curves can be collapsed in the same way as with the experimental results. The length scale l_s should scale with the average length scale of the hierarchies weighted according to their basic eddy row population

$$\text{i.e.} \qquad l_s \sim \sum_{n=1}^{N_o} f(n) \, y_o 2^{n-1} \tag{14}$$

Equation (13) then becomes

$$\frac{\Phi(k_1 l_s)}{U_o^2} \sim \sum_{n=1}^{N_o} \frac{f(n)}{k_1 l_s} \left\{ \exp(-2k_1 l_s R \, 2^{n-1}) - \exp\left(-2k_1 l_s R \frac{y_1}{y_o} 2^{n-1}\right) \right\} \tag{15}$$

where $R = y_o/l_s = 1/\left(\sum_{n=1}^{N_o} f(n) \, 2^{n-1} \right) \tag{16}$

and $\Phi(k_1 l_s)$ is now the energy per unit nondimensional wavenumber $k_1 l_s$. Figure 9(c) shows how these results collapse. Furthermore

$$\int_0^\infty \Phi(k_1 l_s) \, d(k_1 l_s) = \text{constant}$$

invarient with x or t^* which is consistent with the self-preserving flow result. Also l_s grows linearly with t^* and hence x as the computed results show in figure 10. This is also consistent with the self preserving flow result.

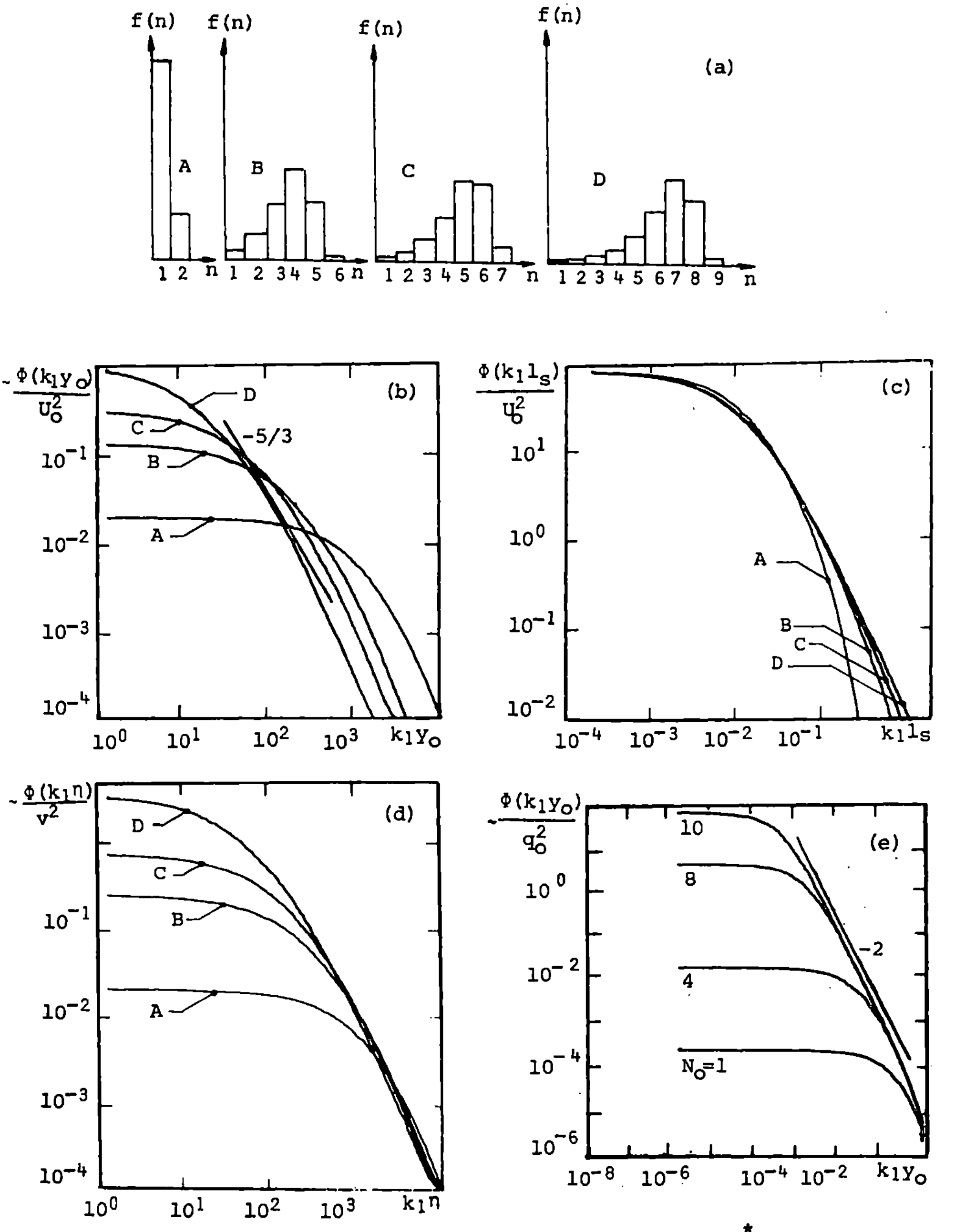

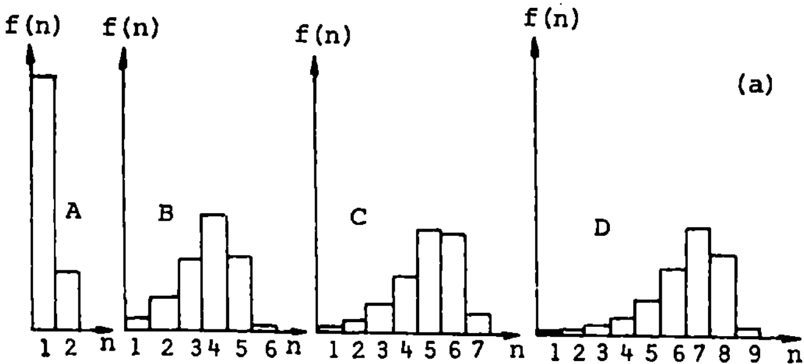

Figure 9(a) Migration of eddies in hierarchy space. t^* ratios are approximately x ratios in experiment. (b) Theoretical spectra according to equations (8) to (12). (c) Same spectra collapsed using U_0 and l_s. (d) Same spectra collapsed using Kolmogorov scaling. (e) Theoretical spectra assuming $f(n) \sim 2^{n-1}$ for $n = 1$ to $n = N_0$ collapsed using velocity scale q_0 and length scale y_0.

117

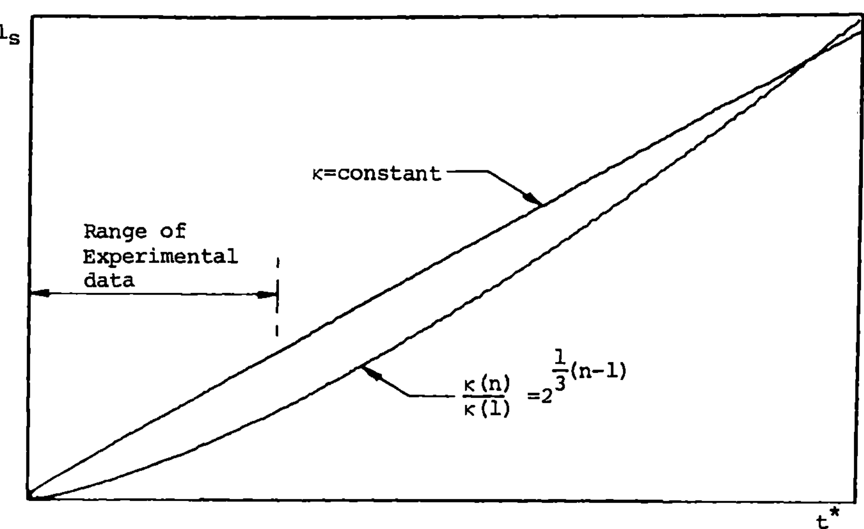

Figure 10. Theoretical growth rate according to equation (14).
Scale factor for axes are arbitrary.

The authors have attempted to modify equation (15) so as to produce the
-5/3 result asymptotically for $t^* \to \infty$. This can be achieved by putting

$$\kappa(n)/\kappa(1) = 2^{\frac{1}{3}(n-1)}$$

in equation (5) and rederiving a new migration equation.
However, there is no obvious physical reason for doing this and the resulting
growth rate of l_s with t^* is not as linear as the simple result with
κ = constant. This is shown in figure 10.
Figure 9(d) shows the theoretical spectra collapse using to the

Kolmogorov length $\qquad \eta \ (= \epsilon^{-\frac{1}{4}} \nu^{\frac{3}{4}})\qquad$ and the

velocity scale $\qquad v \ (= (\epsilon\nu)^{\frac{1}{4}})$

assuming the dissipation $\qquad \epsilon \sim \dfrac{\nu}{y_o^2} \displaystyle\int_0^\infty (k_1 y_o)^2 \Phi(k_1 y_o) d(k_1 y_o)$

and y_o is assumed to be invarient with x.
The collapse is fair even though the curves were obtained from an
entirely different model from that of Kolmogorov. Figure 8 shows a
comparison of the theoretical spectra with experiment for four values of
t^*. These values were used for computing the hierarchy migration in figure
9(a) and the spectra given in figures 9(b) to 9(d). The ratios of the t^*
values chosen are the same as the x station ratios given in figure 5(c).
Presumably, these experimental results would also collapse using the
Kolmogorov length and velocity scale. According to theory, a better high
wavenumber collapse would occur if we use a velocity scale $q_o = U_o \sqrt{f(1)}$ and the

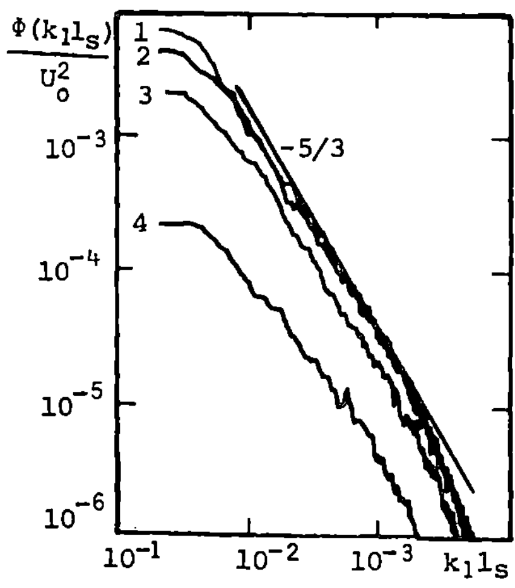

Figure 11. Various spectra taken across flow at station D. (x = 2.5m) 1, z = .004m; 2, z = .064m; 3, z = .144m; 4, z = .224m; l_s = .162m.

length scale y_0. The velocity scale, when squared, scales with the contribution the first hierarchy makes to the total broad-band energy. This collapse is shown in figure 9(e). Figure 11 shows how the spectra vary as we move across the flow. Although there are uncertainties with convection velocities, it is seen that the number of observed hierarchies is approximately constant (unlike wall turbulence) until we are near the edge of the shear layer. This shows that all hierarchies are stretched across the flow.

4. CONCLUSIONS AND DISCUSSION

A number of migration "games" have been attempted by the authors with different assumptions being made for equations (4) and (5). The assumptions presented here are the only ones found so far which give energy production scaling with energy dissipation as well as the other self-preserving flow results. This comes from further analysis.

The model thus represents the simplest possible migration strategy which satisfies the self-preserving flow constraints. The conclusion is that if vortex pairing is the process by which the hierarchies form in wall turbulence then by using the same pairing rules for the plane mixing layer the spectral results follow. The major differences in the two flow situations arise from the different boundary conditions.

In the case of wall turbulence, a continual supply of vorticity is generated at the wall, giving a continual birth of new "basic eddy rows". The stretching and deformation of the eddies leads to eddy "death" by diffusion and vorticity cancellation. This special deformation is due to the presence of a boundary with image vortices as shown by the Biot-Savart law calculation of Perry and Chong (1982). Such a boundary is absent in the plane mixing layer. The hierarchy of eddy scales necessary for obtaining the correct mean flow and Reynolds shear stress field in the wall region could come from a vortex pairing process although another plausible scenario has also been treated in detail in Perry and Chong (1982). The number of basic eddy rows in each hierarchy is constant and eddy row death balances eddy row birth. The scaling of the smallest hierarchy at the wall is governed by the viscous Kline scaling.

In the case of the plane mixing layer, all vorticity is generated at the splitter plate. Downstream from the plate no new vorticity can be created. The finest scale motions originate at the trailing edge of the plate. This is the region where the shear strain rate is at a maximum and the resulting motions persist throughout the flow. No eddy row death occurs since self-preserving flow demands conservation of basic eddy rows per unit

length of flow at least for the model chosen. This model consists of
longitudinal vortex pairs which pair or amalgamate to give geometrically
similar vortex pairs of a higher hierarchy. The lateral scale of one
hierarchy and the next is doubled and so also is the circulation. This
pairing process causes the population of the fine scale motions to be
depleted at the expense of the larger scale motions. This causes the
turbulent energy to redistribute itself to give a power spectral density
close to the -5/3 law as opposed to the -1 power law observed in wall
turbulence.

Experiment shows that the number of hierarchies observed as we move
across the flow remains approximately constant in the plane mixing layer as
opposed to wall turbulence where the number of observed hierarchies diminishes
with distance from the wall. Thus in plane mixing layer turbulence the
vortices are stretched across the flow. In wall turbulence, the eddy heights
within a hierarchy are limited by eddy death and all eddies are attached to
the wall as in the attached eddy hypothesis of Townsend (1976).

It can be shown from the migration equations that l_s is independent of
y_0 provided the initial streamwise spacing of eddy rows also scales with y_0.
This independence of l_s from y_0 is consistent with the Reynolds number
similarity hypothesis since y_0 would depend on viscosity. As a consequence
it follows with further analysis that the spreading rate of the mixing layer
is given by $l_s/x \sim U_0/U_c$. This result was arrived at by Townsend (1976)
from a different line of reasoning.

Further analysis shows that the model implies that $\dot{y}_0 \sim \nu/U_0$ in analogy
with the Kline scaling in wall turbulence and this represents the high
wavenumber length scale and is invarient with x. It can be shown that the
corresponding velocity scale q_0 needed for a high wavenumber collapse of the
data is $q_0 = U_0\sqrt{f(1)} \sim \varepsilon^{\frac{1}{2}}\nu^{\frac{1}{2}}/U_0$. Assuming an "inertial subrange" one can
show that the asymptotic -2 power law for the spectrum must exist on
dimensional grounds alone. The model actually gives this as well as the non-
asymptotic behaviour for the whole range of wavenumbers by the use of velocity
signatures generated by potential vortices with a characteristic direction.
One could porbably construct an "anisotropic" Kolmogorov theory using a
similar migration model. The difference from the present model would be
$y_0 \sim \nu/q_0$ and $q_0 \sim \nu^{\frac{1}{4}} \varepsilon^{\frac{1}{4}}$ with constants characteristic for mixing layers.
By dimensional reasoning one can show that the asymptotic spectral law in
the inertial subrange is a -5/3 law. However, it can be shown that y_0 needs
to vary weakly i.e. as $x^{\frac{1}{4}}$. The migration strategy would need to be more
complex to accomplish this since this implies the complete depletion of
smaller hierarchy populations. The migration equation used in the paper for
generating an asymptotic -5/3 law did not have the above required strategy
and so it violated the self preserving flow conditions. For the range of
data considered, the authors' model appears to be adequate; the -5/3 law
could well be a "red herring" in plane mixing layers.

No account has been made for the spanwise vortices. Only the longitudinal vortices have been considered. Whether their inclusion would alter the
conclusions drawn here has not yet been considered in detail at this stage.

The authors wish to acknowledge the financial assistance of
the Australian Research Grants Scheme and the Australian
Institute of Nuclear Science and Engineering.

REFERENCES

Bandyopadhyay, P. and Head, M.R. 1979. "Visual investigation of turbulent
     boundary layer structure". Cambridge University Eng'g. Dept. Film.
Batchelor, G.K. 1967. "An introduction to fluid dynamics". Cambridge
     University Press.

Breidenthal, R. 1981. "Structure in turbulent mixing layers and wakes using chemical reaction". J. Fluid Mech., 109, pp.1-24.

Brown, G.L. and Roshko, A. 1974. "On density effects and large structures in turbulent mixing layers". J. Fluid Mech., 64, pp.775.

Head, M.R. and Bandyopadhyay, P. 1981. "New aspects of turbulent boundary layer structure". J. Fluid Mech., 107, pp.297-338.

Kline, S.J. 1967. "Observed structural features in turbulent and transitional boundarylayers". J. Fluid Mech., 95, pp.305.

Lighthill, M.J. 1963. "Laminar boundary layers". (ed. Rosenhead), pp.48-88. Clarendon.

Perry, A.E. and Abell, C.J. 1975. "Scaling laws for pipe flow turbulence". J. Fluid Mech., 67, pp.257-271.

Perry, A.E. and Abell, C.J. 1977. "Asymptotic similarity of turbulence structures in smooth- and rough-wall pipe flow". J. Fluid Mech., 79, pp.785-799.

Perry, A.E. and Chong, M.S. 1981. "A physical interpretation of the spectra of wall turbulence". Third Symp. on Turbulent Shear Flow, University of Calif., Davis. Session 1, pp.15-20.

Perry, A.E. and Watmuff, J.H. 1981. "The phase-averaged large-scale structures in three-dimensional turbulent wakes". J. Fluid Mech., 103, pp.33-51.

Perry, A.E., Lim, T.T. and Teh, E.W. 1981. "A visual study of turbulent spots". J. Fluid Mech., 104, pp.285.

Perry, A.E. and Chong, M.S. 1982. "On the mechanism of wall turbulence". J. Fluid Mech., 119, pp.173-218.

Townsend, A.A. 1951a. "On the fine-scale structure of turbulence". Proc. R. Soc. Lond., A209, pp.534-542.

Townsend, A.A. 1951b. "The diffusion of heat spots in isotropic turbulence". Proc. R. Soc. Lond., A209, pp.418-430.

Townsend, A.A. 1976. "The structure of turbulent shear flow". Cambridge University Press.

Winant, C.D. and Browand, F.K. 1974. "Vortex pairing: the mechanism of turbulent mixing-layer growth at moderate Reynolds number". J. Fluid Mech., 63, pp.237.

ON THE RELATION BETWEEN THE
THUNDERSTORM UPDRAFT AND TORNADO FORMATION

by

Richard Rotunno

National Center for Atmospheric Research*
Boulder, Colorado USA

Introduction

Over the past ten years or so, two new sources of data concerning tornadoes have appeared. They come from i) Doppler-radar,used to study the interior of tornadic thunderstorms [1] and ii) systematic surveillance of the tornado and its surroundings by ground-based observers [2]. Important new information concerning the life-cycle and position of the tornado in respect to the parent thunderstorm has been obtained. A synthesis of these results may be found in the paper by Lemon and Doswell [3] and is briefly summarized here. Salient features are that

i)   tornadoes reach their most violent stage during the <u>declining</u> phase of the thunderstorm,

ii) during this time, the center of rotation (detected by the Doppler radar) moves from the updraft maximum to a location between the updraft and the downdraft which is intensifying,

iii) since the National Severe Storms Laboratory has begun its systematic surveillance of tornadoes (1971), no tornado has been found to exist longer than 40 minutes [4] and the typical intense tornado exists for 20-30 minutes [5],

iv) visual observations indicate the tornado, in its intense phase is on or close to the location which separates the thunderstorm inflow (updraft) and outflow (downdraft)[6].

The first three of these points are illustrated by Fig. 1 taken from [7]. Just before the tornado appears (1543) there is a broad area of updraft with the center of circulation just to the south of the updraft maximum. Before the time the next radar scan of the thunderstorm was completed (1553), a tornado was observed to form. What was formerly a broad updraft now is filled with downdraft and the updraft has a contorted shape. The tornado was observed at the location marked in the figure; this location is near the boundary between the updraft and the downdraft. The majority of these systems (termed 'mesocyclones') decay after this change in structure takes place although approximately one-quarter of those observed redevelop (as did this one), and subsequently repeat the process whereby the downdraft intensifies and (another) tornado develops [5].

Figure 2 is a photograph of a tornado which was produced by the same storm system about an hour earlier. The view is to the northwest; the clear air to west-southwest (termed the 'clear-slot' by Lemon and Doswell [3])is indicative of downdraft air. The visible funnel is in updraft air, but as time passes, becomes engulfed by cool downdraft air and ultimately decays.

*The National Center for Atmospheric Research is sponsored by the National Science Foundation.

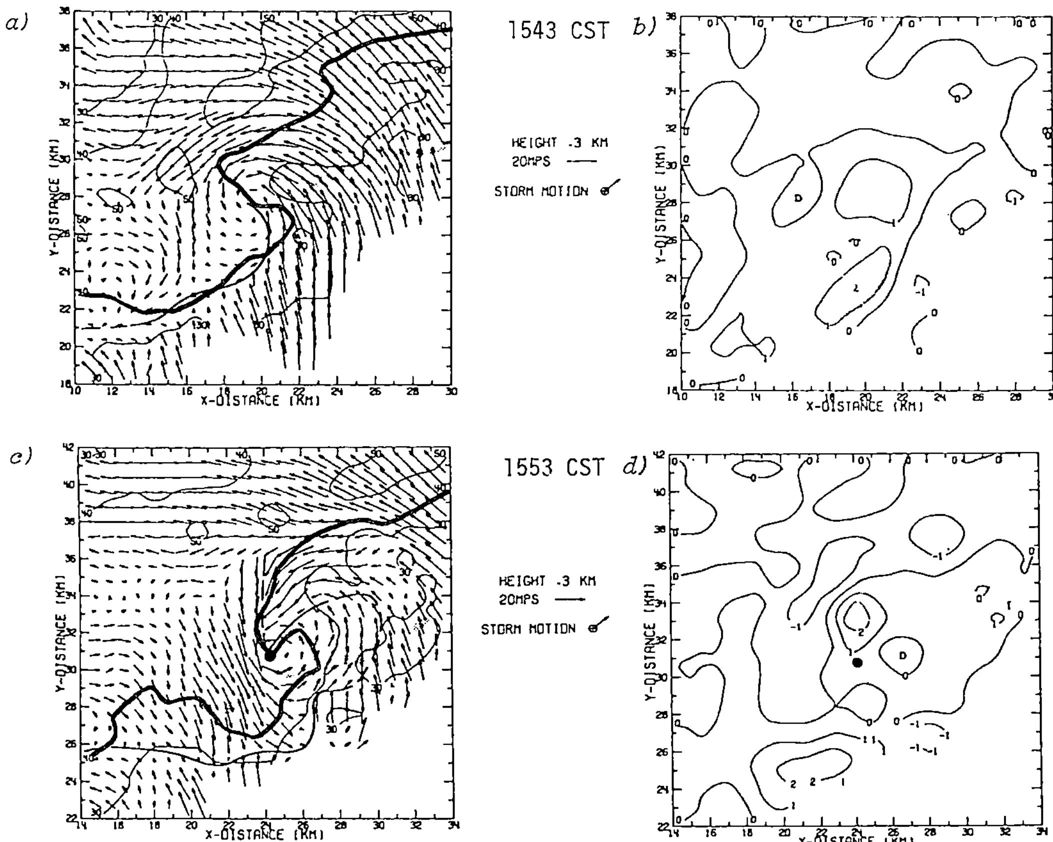

Figure 1. Horizontal storm relative winds (with radar reflectivity (dBZ) superimposed, heavy solid line is the 40dBZ contour) at a)1543 CST (appearance of incipient tornado vortex) and c) 1553 CST (tornado on the ground) and vertical velocity field at b) 1543 CST and d) 1553 CST at z=300 m above ground level. The heavy dot indicates the position of a strong shear anomaly in the raw radar data.

So, in summary, there is significant change in the updraft structure which accompanies tornado formation.

An explanation for this change in updraft structure was proffered by Lemon <u>et. al.</u> [8]. Now, at the time that paper was written <u>multiple</u> Doppler radar was not available and changes in updraft structure were inferred from changes in the reflectivity patterns. A relatively echo-free region generally indicates an updraft so powerful that precipitation hasn't time to form [9]. Lemon <u>et. al.</u> [8] noted that the echo-free region typically begins to fill with echo as the mesocyclone circulation (inferred from <u>single</u> Doppler measurements) increases. So, Lemon <u>et. al.</u>'s [8] hypothesis is that the increasing mesocyclone circulation effectively 'chokes' the updraft in analogy to the fluidic vortex valve [10].

This explanation is largely substantiated by [7]. Figure 3 taken from that paper shows the flow fields azimuthally-averaged about the circulation center. At 1543, there is a broad updraft at 1553 the swirling motion increases at low levels and decreases at higher levels and the updraft now contains a core of downward motion.

Figure 2. One of the Oklahoma City tornadoes of 8 June 1974. The view is from the southeast. As discussed in [3], the "clear" slot is associated with the downdraft to the left of the tornado while the updraft-associated lowered cloud base is to the right. (Photo from Burgess et.al. [29])

The relationship between the change in updraft structure and tornado formation is less clear owing primarily to the lack of data on the tornado scale $[0(10^2m)]$. However, it appears from Fig. 1 that the tornado is at one of the foci of an elliptically shaped mesocyclone. Further, Fujita [30] adumbrated and Agee et.al. [31] more fully developed the idea that tornadoes are asymmetrically located with respect to their parent circulations by examination of surface damage patterns.

In this lecture, we would like to develop Lemon et.al's [8] argument further and to then show how the change in updraft structure may lead to tornado formation. This will be done using a simple three-dimensional numerical model (to some this may appear an oxymoron) of a mesocyclone. The numerical model is based upon Ward's [11] laboratory model of a mesocyclone. The results may be summarized as follows. Given a non-rotating updraft subsequent addition of swirling motion over the lowermost portion of inflow induces an adverse pressure gradient which modifies the updraft structure so that it will contain a central downdraft. The resulting flow is unstable to three-dimensional perturbations; asymmetries form which develop into smaller circulations imbedded within the larger circulation. This lends support to the hypothesis that the tornado results through the instability of the mesocyclone [7].

The Model

Figure 4 is a schematic of the Purdue vortex generator based on Ward's [11] design. The basic physical model upon which Ward's model and, ipso facto the present model is based has been discussed at length by Davies-Jones [12] Church et.al. [13] and most recently by Snow [14]. There is a cylindrical chamber bounded above by a baffle over which suction is created by an exhaust fan. The rotating screen imports an amount of angular momentum to the in-rushing air. The basic physical mechanism is that low-level convergence of air which conserves its angular momentum leads to large azimuthal velocities.

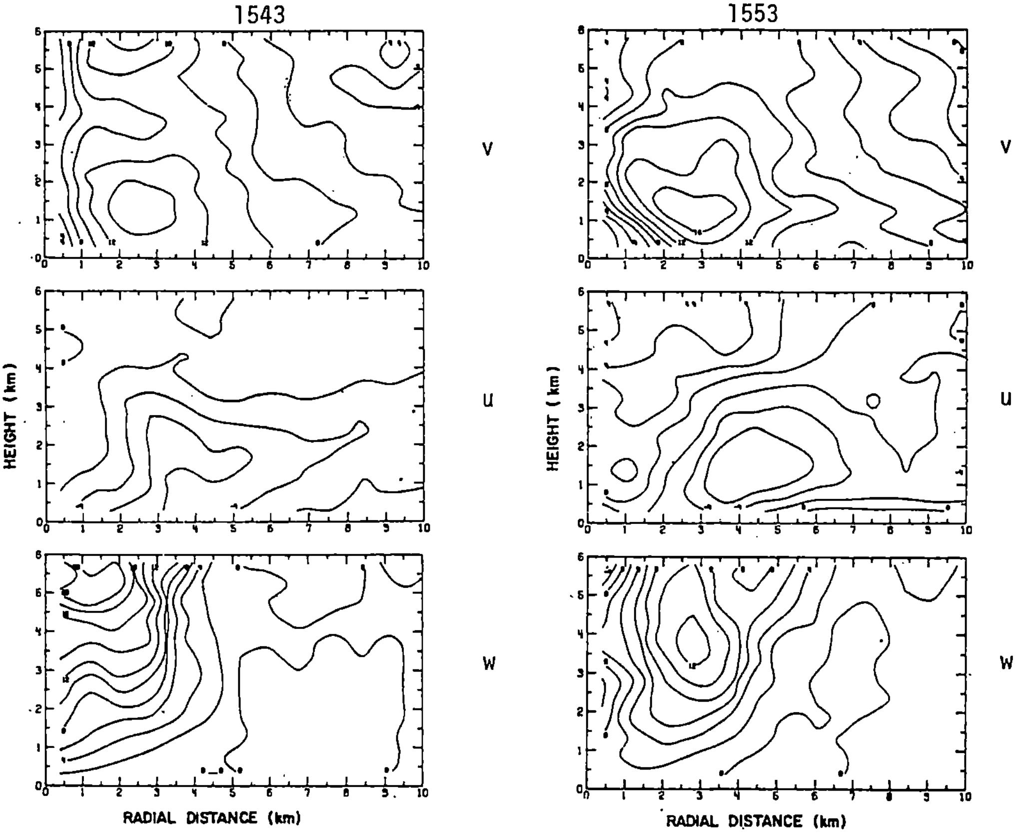

1543   1553

Figure 3. Radial-height distributions of mean tangential (v), radial (u) and vertical (w) wind during a) apperance of incipient tornado vortex (1543 CST) and b) when tornado is on the ground (1553 CST). Positive velocities (m s^{-1}) designate cyclonic, outward and upward motion. Radial distances are from the mesocirculation vertical axis [7].

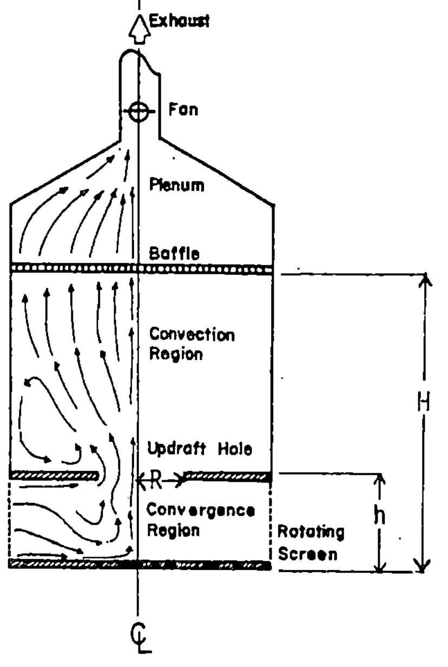

Figure 4. Schematic of the Ward-type vortex generator/tornado simulator, showing the major components of the apparatus [13].

Among the various non-dimensional numbers which might govern the flow in the vortex generator one has been found to be by far the most important. It is the swirl ratio

$$S \equiv \frac{R\Gamma_R}{2Q} \tag{1}$$

where R is the radius of the updraft hole, Γ_R is the circulation at the outer radius of the convergence region, h is the depth of the convergence region and Q is the volume flow rate. Another important parameter is the radial Reynolds number,

$$R_e \equiv \frac{Q}{2\nu} \tag{2}$$

where ν is the kinematic viscosity.

The flow in the inner cylindrical region bounded by the radius R and the height of the baffle, H is studied herein. The equations governing the flow in this domain are the momentum equations

$$r\text{-dir} \qquad \frac{du}{dt} - \frac{v^2}{r} = -\frac{\partial P}{\partial r} + \frac{1}{R_e}\left(\frac{\partial^2 u}{\partial z^2} + \frac{1}{r^2}\frac{\partial^2 u}{\partial \theta^2} - \frac{\partial^2 w}{\partial r \partial z} - \frac{1}{r^2}\frac{\partial}{\partial r}\, r\, \frac{\partial v}{\partial \theta} \right) \tag{3}$$

$$\theta\text{-dir} \qquad \frac{dv}{dt} + \frac{uv}{r} = -\frac{1}{r}\frac{\partial P}{\partial \theta} + \frac{1}{R_e}\left(\frac{\partial}{\partial r}\frac{1}{r}\frac{\partial}{\partial r}\, rv + \frac{\partial^2 v}{\partial z^2} - \frac{\partial}{\partial r}\frac{1}{r}\frac{\partial u}{\partial \theta} - \frac{1}{r}\frac{\partial^2 w}{\partial \theta \partial z} \right) \tag{4}$$

$$z\text{-dir} \qquad \frac{dw}{dt} = -\frac{\partial P}{\partial z} + \frac{1}{R_e}\left(\frac{1}{r}\frac{\partial}{\partial r}\, r\, \frac{\partial w}{\partial r} + \frac{1}{r^2}\frac{\partial^2 w}{\partial \theta^2} - \frac{1}{r}\frac{\partial^2 v}{\partial r \partial \theta} - \frac{1}{r}\frac{\partial}{\partial r}\, r\frac{\partial u}{\partial z} \right) \tag{5}$$

and continuity,

$$\frac{1}{r}\frac{\partial}{\partial r}\, (ru) + \frac{1}{r}\frac{\partial v}{\partial \theta} + \frac{\partial w}{\partial z} = 0 \tag{6}$$

where

$$(u,v,w) = \frac{1}{u_R}(u',v',w'); \quad P = \frac{P'}{\rho u_R^2}; \quad (z,r) = \frac{1}{R}(z',r'); \quad t = \frac{u_R t'}{R};$$

u', v', w' are the dimensional components of the velocity vector in the r', θ, z' directions, respectively; P' is the pressure, ρ the density and u_R is the radial velocity at the edge of the updraft hole.

The general approach toward the numerical solution of Eq. (3)-(6) is to specify u, v, w and P at an initial time and then to "march" Eq. (3)-(5) forward in time to obtain new values for u, v, w. The divergence of Eq.(3)-(5) yields a Poisson equation for P which involves no time derivatives of the dependent variables, hence the new values of u, v, and w may be used to calculate the corresponding new value of P. This process may, in principle, be repeated indefinitely. The details of the numerical procedures are described by Rotunno and Lilly [15]. To complete the position of the problem, boundary conditions are necessary; they are as follows.

126

a) z=0. At the lower surface it is required that no flow leave or enter the domain, hence

$$w(r,\theta,o,t) = 0 \qquad\qquad (8)$$

The condition on P (needed to solve the Poisson equation) is obtained by noting that the LHS of Eq. (5) vanishes at z=0, so, $\partial P/\partial z$ is known because it is equal to terms which involve only spatial derivatives of known quantities. The conditions on u and v are less straight-forward. It is known that a viscous fluid adheres to solid bounding surfaces so that if the main body of the fluid and its boundary are in relative motion a boundary layer over which the velocity changes from that of the boundary to that of the fluid will form. This boundary layer will typically be thin and thus hard to resolve with the limited resolution available to a three-dimensional model. As shown by Rotunno [16], the formation of a cylindrical shear layer/ hollow core vortex can be obtained using a free-slip lower boundary condition. Hence,

$$\frac{\partial u}{\partial z}\,(r,\theta,o,t) = \frac{\partial v}{\partial z}\,(r,\theta,o,t) = 0 \qquad\qquad (9)$$

b) z=H/R. Figure 4 shows there is a baffle through which the air must flow before leaving the lower chamber. We believe that the main effect of the baffle is to eliminate horizontal air motion, that is

$$u(r,\theta,\tfrac{H}{R},t) = v(r,\theta,\tfrac{H}{R},t) = 0 \qquad\qquad (10)$$

These boundary conditions are the cause of some numerical reflections into '2Δx' waves which, in turn, are suppressed by applying a filter to the uppermost four grid points; these uppermost grid points are considered, therefore, the numerical equivalent of the baffle. Since u and v are known at H, w is obtained through Eq. (6). A condition on P is obtained by integrating Eq. (3) from o to r at z=H/R,

$$P(r,\theta,\tfrac{H}{R},t) = P(o,\theta,\tfrac{H}{R},t) - \int_{o}^{r} \left\{ w\,\frac{\partial u}{\partial z} - \frac{1}{R_e}\,\frac{\partial^2 u}{\partial z^2} \right\}_{z=\frac{H}{R}} dr \qquad (11)$$

Because the absolute value of pressure is irrelevant we arbitrarily set $P(o,\theta,\tfrac{H}{R},t) = 0$.

c) r=1. The edge of the <u>updraft hole</u> is modelled as follows. The radial velocity is specified constant over the depth of the <u>convergence region</u>, h. A fictitious wall is placed between h and H where the radial velocity is set to zero. Thus

$$u(1,\theta,z,t) = \begin{cases} 1 & 0 \le z \le \dfrac{h}{R} \\[2mm] 0 & \dfrac{h}{R} < z \le \dfrac{H}{R} \end{cases} \qquad (12)$$

A condition on w is obtained by requiring there be zero azimuthal component of vorticity at r=1, viz.

$$\left.\frac{\partial w}{\partial r}\right|_{r=1} = \left.\frac{\partial u}{\partial z}\right|_{r=1} \qquad\qquad (13)$$

The justification for this is that there are no sources of azimuthal vorticity
in the convergence region upstream of the <u>updraft hole</u> save for the thin
boundary layers on the upper and lower plates. Thus, the air possesses zero
azimuthal vorticity at r=1 because it was zero to start with and encounters
no sources enroute.

The azimuthal velocity is specified at r=1 as

$$v(1,\theta,z,t) = \frac{v_R}{u_R} [1 - \exp(-2t)] \quad 0 \leq z \leq \frac{h}{R} \tag{14}$$

to simulate the spin-up of the screen to a constant rotation rate. That the
condition (14) sets the swirl ratio for the calculation may be seen by
noting that $Q = Ru_R h$, $\Gamma_R = Rv_R$ so that by Eq. (1) one obtains

$$S = \frac{R}{2h} \frac{v_R}{u_R} . \tag{15}$$

For all the calculations presented here we set R=2h so that the swirl ratio
$S = v_R/u_R$. Along the fictitious wall ($h/R \leq z \leq H/R$) we require zero vertical vorticity.

$$\frac{\partial}{\partial r} (rv) \bigg|_{r=1} = 0 \tag{16}$$

The boundary condition on $\partial P/\partial r \big|_{r=1}$ is obtained from Eq. (3).

    d) r=0. There is in principle no need to require anything of the flow
in the middle of the domain. However, certain practical considerations
dictate the use of a small inner cylinder which does not exist in the
laboratory. The system of cylindrical coordinates, in which the code is
written, has ever decreasing azimuthal grid lengths $r\Delta\theta$ as $r \to o$. The
condition for viscous stability that $\Delta t < R_e \Delta^2/4$ is thus very expensive to
satisfy. Because 1) most of the interesting dynamics take place in this
problem about halfway between r=0 and r=1 and 2)We desired to keep the code
as simple as possible. The placement of a small inner cylinder at r=0 was
deemed a reasonable compromise. Then conditions on this cylinder are that
there be zero radial motion

$$u(a,z,\theta,t) = 0 \tag{17}$$

zero azimuthal shear stress

$$\frac{\partial}{\partial r} \left(\frac{v}{r}\right) \bigg|_{r=a} = 0 \tag{18}$$

and zero vertical shear stress

$$\frac{\partial w}{\partial r} \bigg|_{r=a} = 0 \tag{19}$$

The condition on $\partial P/\partial r \big|_{r=a}$ is then obtained from Eq. (3). I take a=.02 for
the present work.

The parameter values for this experiment are S=1, R_e=150, HR^{-1}=.5, a=.02. The Reynolds number is to be interpreted as a turbulent one based on an eddy viscosity, thus is much lower than the values given by Church et.al. [13] based on the kinematic viscosity of air at room temperature.

<u>Results</u>

a. Axisymmetric regime

The flow at t=0 is determined by the solution to the system

$$\frac{\partial}{\partial r} \left(\frac{1}{r} \frac{\partial \psi}{\partial r}\right) + \frac{1}{r} \frac{\partial^2 \psi}{\partial z^2} = 0 \qquad (20)$$

$$\psi(r,o) = 0; \quad \psi(a,o) = 0; \quad \partial\psi/\partial z(r,\tfrac{H}{R}) = 0; \quad \psi(1,z) = R\int_o^z u(1,z,\theta,t)dz$$

where $u(1,z,\theta,t)$ is given by Eq. (12). The LHS of Eq. (20) is the azimuthal vorticity and $\psi(r,z)$ is the streamfunction. The solution to Eq. (20) then yields a state of axisymmetric irrotational flow as obtained by Rotunno [16, Fig. 4.1]. Figure 5 displays the velocity and pressure fields associated with the solution to Eq. (20). The radial velocity indicates radial inflow monotonically increasing to zero at the center axis (r=a) and at the top at z=2. The vertical velocity increases upward ($0 \leq z \leq h/2$) as demanded by continuity given the radial variation of u. There is a maximum w in the vicinity of (r=1, z=1/2) which is the edge of the <u>updraft hole</u>. This behavior is demanded by the conditions (12) and (13) which require $\partial w/\partial r$ to be large and positive at r=1, z=1/2. The pressure, then, by Bernoulli's theorem for irrotational flow reflects the velocity field; there is low pressure near r=1, z=0. This solution is, of course, a steady solution to the fully non-linear system Eqs. (3)-(6).

As tangential velocity is transported inward by the radial motion, substantial changes occur in the updraft structure. Figure 6 displays the velocity and pressure fields at t=3.8. The tangential velocity exhibits a pronounced maximum near r=.4, z=0 and decays with height. The pressure field on the lower surface has a high pressure ring surrounding a region of low pressure on the central axis. The vertical velocity field now contains a central core of downdraft. Beyond a radius of .75 or so the u, w and P fields are similar to the initial state described above. One may understand the high pressure ring to be a consequence of the acceleration of radial motion field at radii near r=1. Equation (3) written at the lower surface, neglecting friction, is

$$\frac{\partial P}{\partial r} = -\frac{\partial}{\partial r}\frac{u^2}{2} + \frac{v^2}{r} \qquad (21)$$

As discussed by Rotunno [16,17], at large radii, the first term dominates the second and is negative. Hence the pressure increases inward. At smaller radii the second term dominates (cyclostrophic balance) and is necessarily positive, so the pressure decreases inward. Cyclostrophic balance is a fair approximation within the radius of the high pressure ring. Taking a vertical derivative of the cyclostrophic equation obtains

$$\frac{\partial}{\partial z} \left(\frac{\partial P}{\partial r}\right) = \frac{1}{r} \frac{\partial v^2}{\partial z} \qquad (22)$$

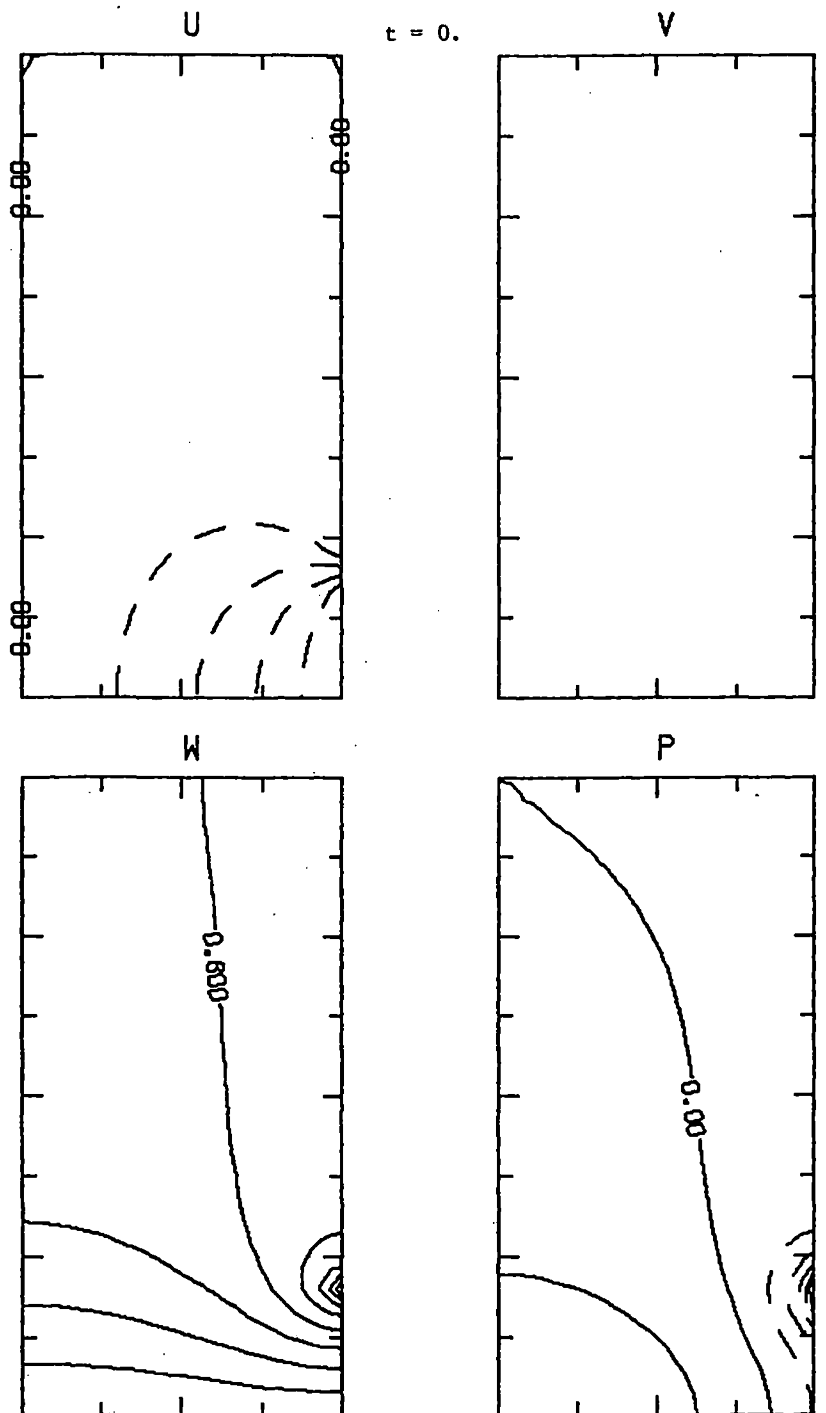

Figure 5. Initial state of velocity and pressure fields. Velocity and pressure contour increments = .2.

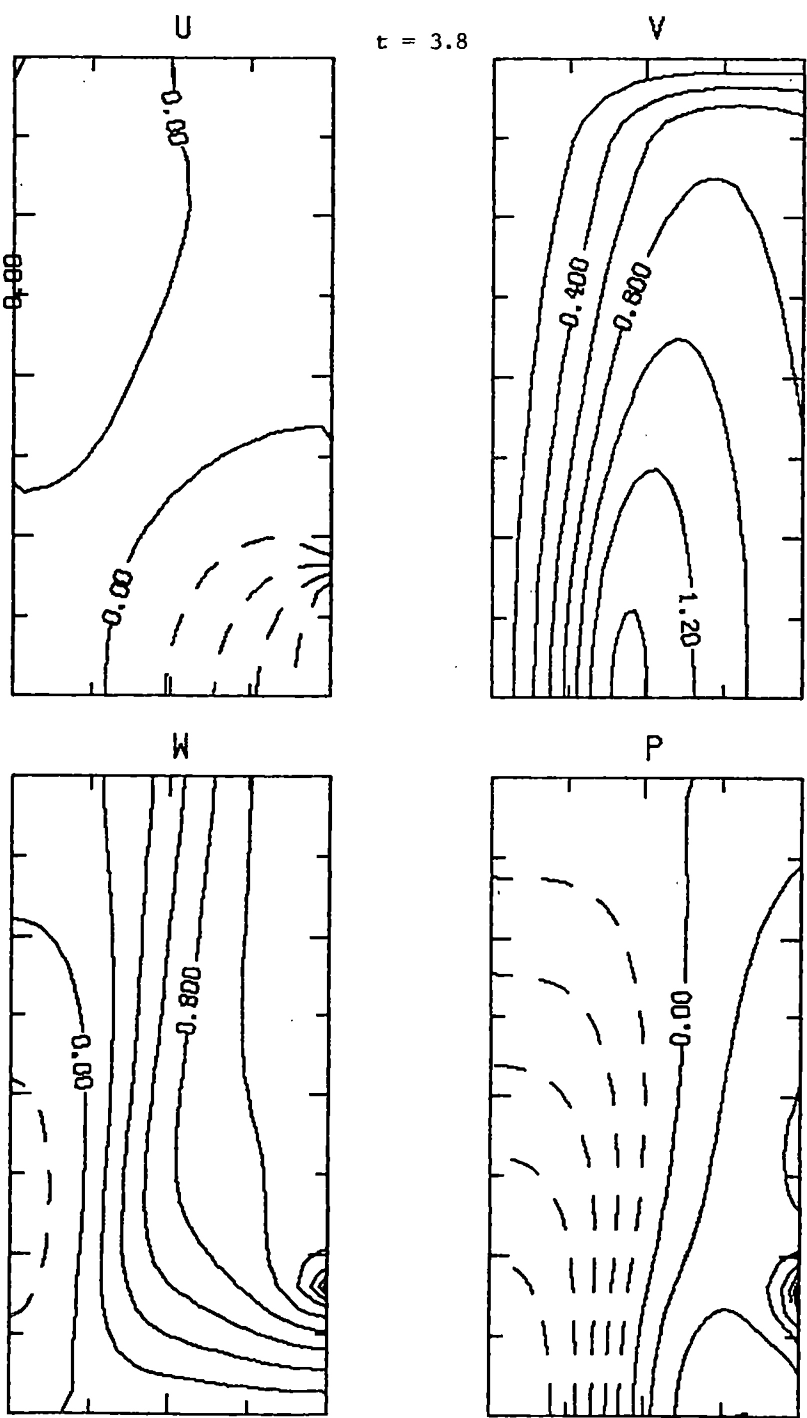

Figure 6. Velocity and pressure fields (t = 3.8). Note formation of central downdraft.

Figure 6 shows that the RHS is negative and thus the radial pressure gradient
decreases with height. Since the pressure at large radius suffers little
change with height, Eq. (22) implies that the pressure on the central axis
must increase with height. So there is an adverse vertical pressure gradient
on the central axis this gives rise to the central downdraft.

b.   Aside: the steady-state axisymmetric flow

Now unlike the thunderstorm updraft discussed in the introduction, the
present model reaches a steady-state owing to the fact that the volume flow
rate is held fixed through the boundary condition, Eq. (12). This steady-
state configuration (t=10) is displayed in Fig. 7; here the central downdraft originates at the top of the domain and penetrates to the lower
surface. It is evident in both Fig. 6 and 7 that the radial velocity
becomes zero at an intermediate radius which implies that the air stream
originating in the lower right of the domain separates from the lower surface and then continues upward. The radius at which this occurs $[u(r_c,o)=0]$
is termed the <u>core radius</u>. The core is being filled with non-swirling air
through the upper boundary and so v exhibits a nearly cylindrical shear
layer which separates the inner stagnate core from the outer flow. The
magnitude of the updraft has increased to ~ 1.7 at r=1, z=2, which reflects
the fact that the same volume flow rate is imposed upon a smaller area
owing to the flow separation. This acceleration of the vertical velocity
along the outer edge of the domain induces a lowering of pressure with
height at that radius. And, although Eq. (22) still correctly predicts
that the radial pressure gradient decreases with height, the rapid decrease
with height of the outer pressure requires that the central pressure
decrease slightly with height.

The determination of the core radius as a function of the external
parameters has been studied in the context of other applications [18] and
[10]. The approach there is to assume a two-regime flow -- an inner
cylinder of stagnant flow separated from an outer irrotational flow --
then to calculate the pressure drop across the core at the exhaust (in the
schematic in Fig. 4 this corresponds to z'=h, o < r'<R) in terms of the
imposed circulation and volume flow rate. By maximimizing the volume flow
rate (for a constant pressure drop [18]) or minimizing the pressure drop
(for a constant volume flow rate [19]) with respect to the core radius, a
relation between r_c and the swirl ration, S is obtained. Davies-Jones [19]
found that although this procedure gives the correct asymptotic behavior
($\lim_{s\to o} r_c(S)=0$; $\lim_{s\to\infty} r_c(S)=1$), the theory generally predicts core radii larger
than those observed. Baker and Church [20] were able to improve the agree-
ment between theory and experiment by using a more complicated flow model
and, in a crude way, accounting for the effects of turbulence in the core.
The following paragraphs will discuss a simple theory for $r_c(S)$. Since
this is only peripherally related to the specific theme of this lecture,
the reader may continue at section c without loss of continuity.

Eqs. (3)-(6) may be written as

$$\nabla H = \underset{\sim}{v} \times \underset{\sim}{\omega} - \frac{1}{R_e} \underset{\sim}{\nabla} \times \underset{\sim}{\omega} \tag{23}$$

where $H \equiv P + 1/2\, \underset{\sim}{v} \cdot \underset{\sim}{v}$ is the total head and $\omega = (\xi,\eta,\rho)$ is the vorticity
vector. Now consider the integral of Eq. (23) around the domain ($0 \leq r \leq 1$,
$0 \leq z \leq 2$); neglecting friction one obtains

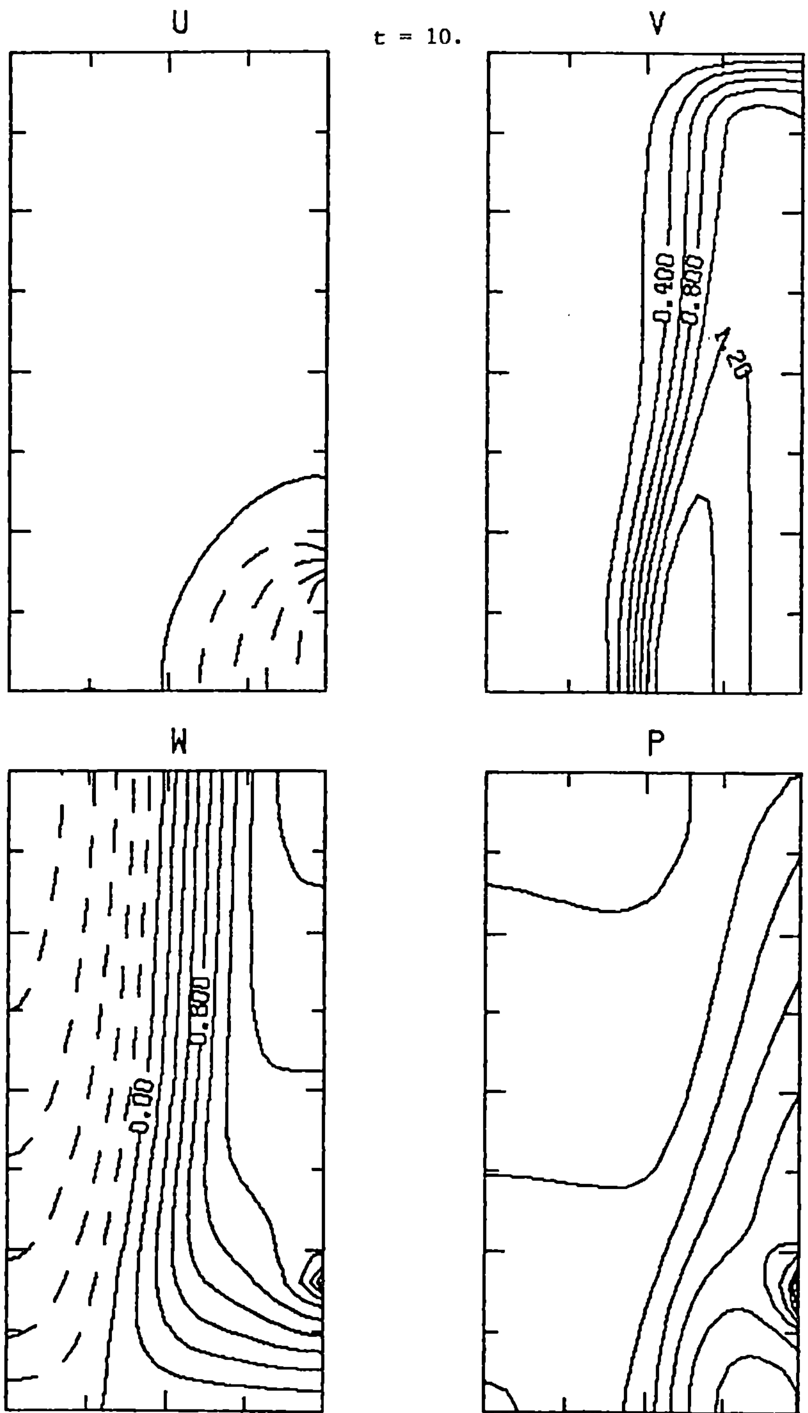

Figure 7. Steady-state velocity and pressure fields (t = 10.0).

$$\oint (\underset{\sim}{v} \times \underset{\sim}{\omega}) \cdot d\underset{\sim}{\ell} = 0 \tag{24}$$

or

$$\int_{r=0}^{1} [(v\zeta-w\eta)\Big|_{z=0} - (v\zeta-w\eta)\Big|_{z=2}]dr + \int_{z=0}^{2} [(u\eta-v\xi)\Big|_{r=1} - (u\eta-v\xi)\Big|_{r=0}]dz=0 \tag{25}$$

Using the boundary conditions, Eq. (25) may be simplified to

$$\int_{r=0}^{1} [w\eta\Big|_{z=2} + \frac{\Gamma^2}{r^3}\Big|_{z=0}]dr = 0 \tag{26}$$

Judging from Fig. 7, $\eta\Big|_{z=2} \approx - \partial w/\partial r$; hence Eq. (26) then gives

$$w^2(1,2) - w^2(0,2) = 2 \int_{o}^{1} \frac{\Gamma^2}{r^2}\Big|_{z=0} dr \tag{27}$$

Reverting back to the two-regime flow, we take

$$w(r,2) = \begin{cases} 0 & 0 \leq r < r_c \\[2ex] \dfrac{2Q}{R^2-r_c^2} & r_c < r \leq 1 \end{cases} \tag{28}$$

and

$$\Gamma(r,0) = \begin{cases} 0 & o \leq r < r_c \\[2ex] \Gamma_R & r_c < r \leq 1 \end{cases} \tag{29}$$

which allows Eq. (27) to be written as

$$s^2 = \frac{x^2}{(1-x^2)^3} \tag{30}$$

where $x \equiv r_c/R$. Although $w^2(0,2)$ is not much smaller than $w^2(1,2)$ (it is approximately one-third as large), the prediction of Eq. (30) for S=1 is x=.55 which compares favorably to the model result (x=.50). However, both overestimate the measured value of x=.33 [20, Fig. 2].

We will show in the following section that the axisymmetric flow shown in Fig. 7 is unstable to three-dimensional perturbations. These perturbations grow to finite amplitude and modify the aximuthally-averaged state; in particular the core size is decreased to $x \approx .367$.

c.   Asymmetric motion

Ward [11] noted that for $S > O(1)$ the flow within the vortex chamber becomes asymmetric and is characterized by smaller subsidiary vortices which he called multiple vortices. Ward hypothesized that these vortices form as the result of the instability of the cylindrical shear layer associated with the radial variation of $v(r)$ (see his Fig. 20). At this stage we are able to investigate this hypothesis by adding a small amount of random perturbations to the fields shown in Fig. 7 and then integrating further in time to note changes, if any.

The results of the test are affirmative; the flow in Fig. 6 is unstable and asymmetries begin to develop shortly after the addition of random perturbations. After a time, the azimuthal wavenumber m=2 appears to dominate; the unstable waves grow to finite amplitude, while modifying the mean flow, and subsequently reach an equilibrium state where these finite amplitude waves may coexist with a modified mean flow (see Figs. 8 and 9). The horizontal flow pattern of Fig. 9 clearly indicates two smaller centers of rotation at some distance from the center of the domain; the lowest pressures are found in association with these centers. That these centers of rotation tilt-clockwise with height (although embedded in a counter-clockwise flow) is illustrated in Fig.10 where a three-dimensional contour representation of the pressure field is displayed.

Analytical results concerning the linear stability of flows $(u(r,z)$, $v(r,z)$, $w(r,z))$ such as that represented by Fig. 7 (referred to as the base flow) are nonexistent. Some analytical results are available for flows $(0, v(r), w(r))$, however, very few general statements can be made [21]. A few general conditions concerning simpler flows and perturbations thereof are briefly reviewed here.

Rayleigh [22] showed that the flow $(0,v(r),0)$ is stable to two-dimensional $(r-\theta)$ perturbations if the vorticity $r^{-1}\partial(rv)/\partial r$ is a monotonic function of r. Later Rayleigh [23] found the flow $(0,v(r),0)$ is stable to axisymmetric $(r-z)$ perturbations if $r^{-3}\partial(r^2v^2)/\partial r > 0$. Howard and Gupta [21] generalized this criterion for flows $(0,v(r),w(r)$; i.e. the flow is stable to axisymmetric perturbations if $r^{-3}\partial(r^2v^2)/\partial r /(\partial w/\partial r)^2 > 1/4$. No such general criteria exist for three-dimensional perturbations to flows $(0,v(r),w(r))$ even with $w(r)=0$. Hence, the stability of such flows has been investigated on a case-by-case basis. Rather than examining the linear stability of some flow $(0,v(r),w(r))$ which would anyway be a crude approximation to the base flow (owing to the lack of radial motion and vertical variation), we describe some special numerical calculations made to investigate certain aspects of the stability of the base flow.

The stability of the base flow to axisymmetric disturbances is examined by adding random noise to the base flow and integrating forward with the axisymmetric version of the three-dimensional model described in Section 2. Although the Howard and Gupta [21] criterion is violated at most vertical levels, the numerical integrations reveal that the base flow is stable to axisymmetric perturbations.

The idea that MV* arises as an instability of a cylindrical layer of shear in the azimuthal velocity to two-dimensional (independent of z) perturbations has been proposed by Ward [11] Davies-Jones and Kessler [24], Snow [25], Staley and Gall [26]. There are problems with this idea from the outset. Snow [25] analysed the linear stability of a cylindrical vortex

---

*MV = multiple vortex

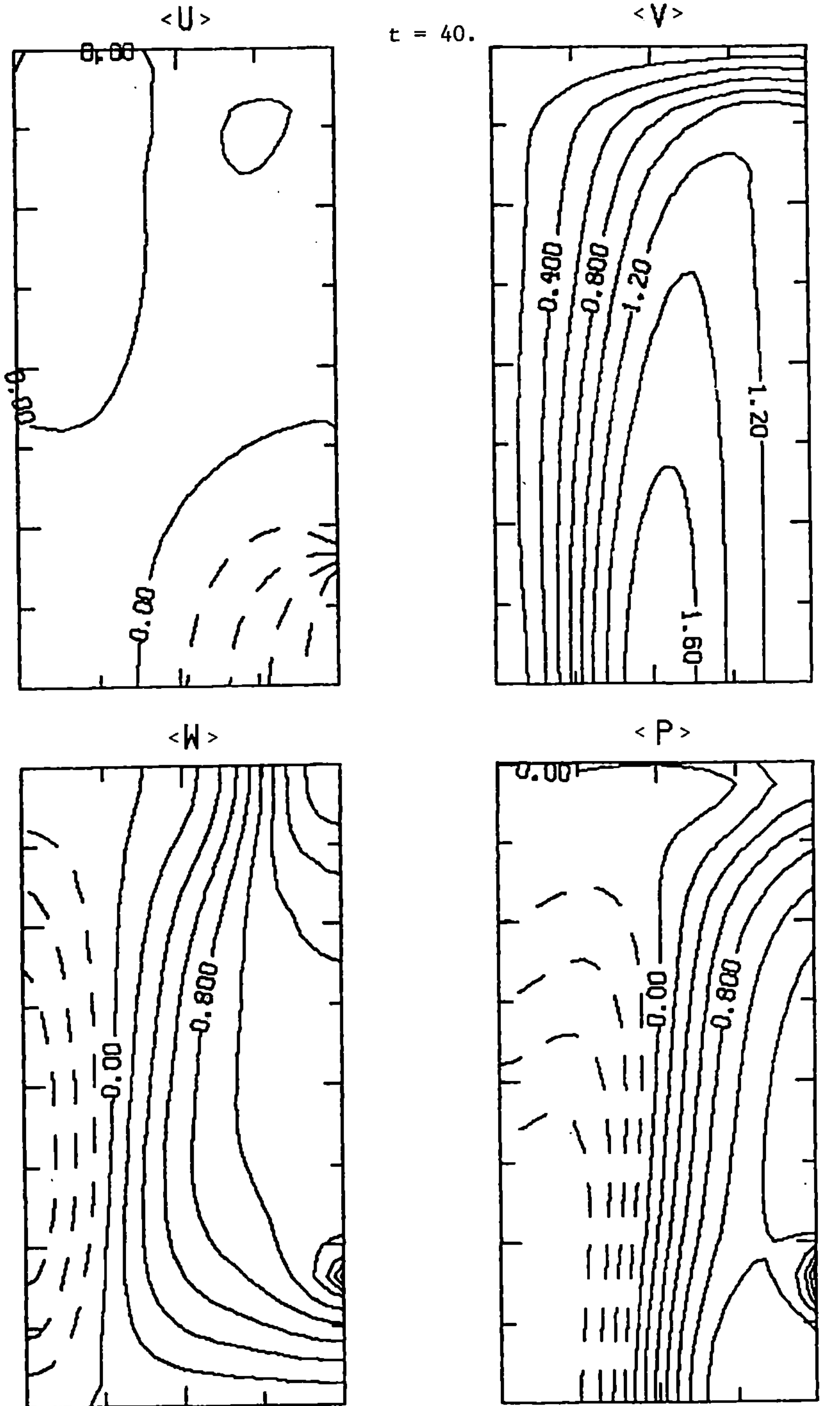

Figure 8. Azimuthally-averaged fields after asymmetries have grown to finite amplitude and have modified the mean flow (t = 40.).

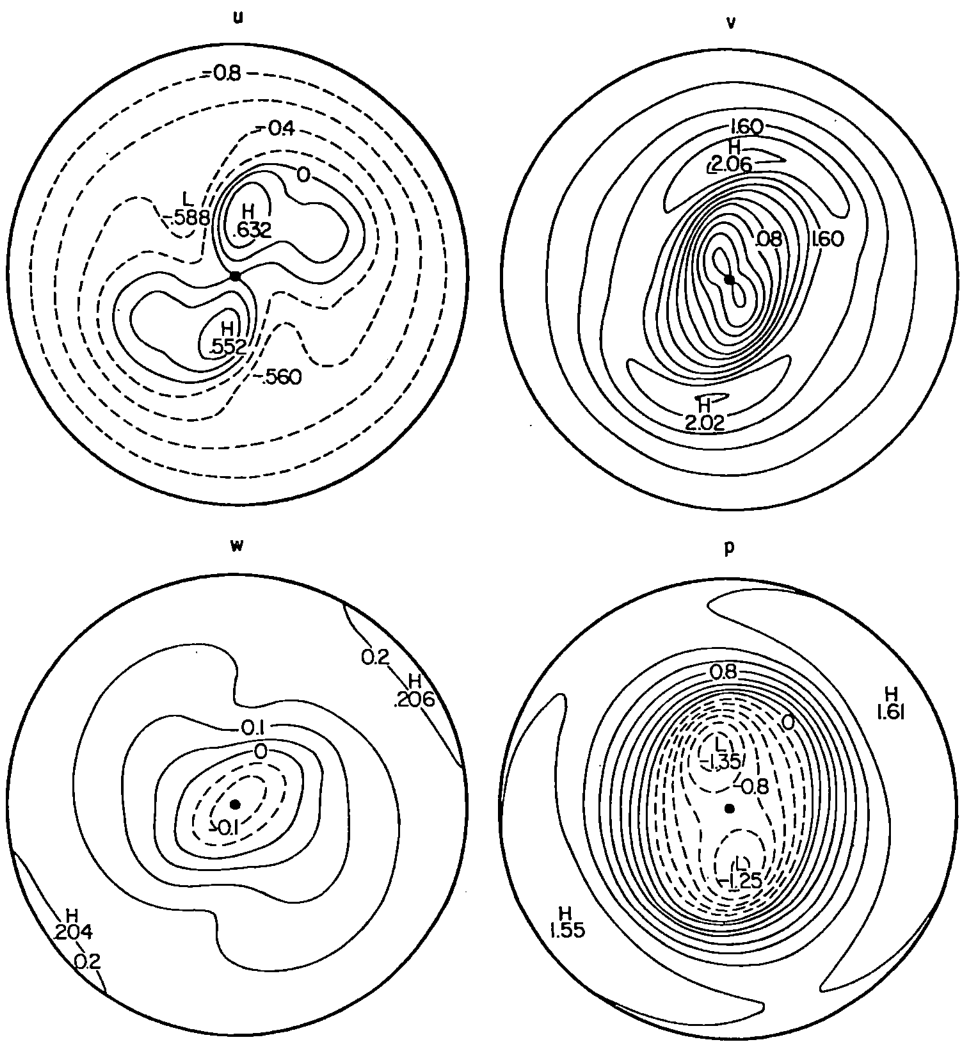

Figure 9. r-θ section of velocity and pressure fields at $z = \Delta z/2$, $t = 40$. Note smaller circulations within large vortex.

sheet of finite thickness and found that no matter how the parameters
(shear layer thickness. etc.) were varied, disturbances with azimuthal
wavenumber m=1,2 are stable. Rotunno [27] came to the same conclusion
for an infinitely thin vortex sheet. Hence, it is impossible for the
2 MV's to be a direct consequence of the instability of the base flow
$(0,v(r),0)$. Now, it may be that in a non-linear two-dimensional model,
azimuthal wavenumber m=2 could gain energy through non-linear transfer.
To test this idea, a two-dimensional version of the three-dimensional
model described in Section 2 using the flow $(0,v(r),0)$ (where $v(r)$ is taken
from the base flow at several different levels) as the initial condition
was run. Again, after the addition of random noise, the ensuing integration gave no evidence of instability.

Hence, the ineluctable conclusion to be drawn from these experiments
is that the instability is an essentially three-dimensional phenomenon.
Next we consider factors which determine the shape of the asymmetric motion.

The radial, azimuthal and axial components of the vorticity vector are
respectively

$$\xi = \frac{1}{r}\frac{\partial w}{\partial \theta} - \frac{\partial v}{\partial z} \qquad \eta = \frac{\partial u}{\partial z} - \frac{\partial w}{\partial r} \qquad \zeta = \frac{1}{r}\frac{\partial}{\partial r}(rv) - \frac{1}{r}\frac{\partial u}{\partial \theta} \qquad (21,a,b,c)$$

Because the base flow is axisymmetric, $\xi = -\partial v/\partial z$ and $\zeta = r^{-1}\partial(rv)/\partial r$. The
distribution of vorticity in the base flow can be deduced from Fig. 7. It
can be roughly characterized as being small everywhere except in the cylindrical layer where w and v increase rapidly with radius and in the <u>baffle</u>
layer near the top where u, v have strong variation with z. Below the
baffle layer and within the cylindrical shear layer the local vorticity
vector points upwards ($\zeta > 0$) in the <u>negative</u> azimuthal direction
($\eta \approx -\partial w/\partial r < 0$) and slightly outward ($\xi$ is slightly positive owing to the
slight decrease of v with z). Consider now the vortex lines associated
with this distribution.

Recall that a <u>vortex line</u> is that <u>curve</u> which is everywhere tangent to
the vorticity field. Hence, based on the rough characterization of the
vorticity distribution given above, the vortex lines of the cylindrical
shear layer are helices of negative pitch angle which expand slightly with
height. Some of the vortex lines* which intersect the maximum value of ζ
at z=0 are shown in Fig. 11.

These lines represent integrations of the equations

$$\frac{dr}{dz} = \frac{\xi(r,z)}{\zeta(r,z)} \qquad \frac{d\theta}{dz} = \frac{1}{r}\frac{\eta(r,z)}{\zeta(r,z)} \qquad (22\ a,b)$$

where ξ, η and ζ are obtained from the model.

---

*There are infinitely many of these vortex lines. Taken together they
define a <u>vortex-sheet</u>, which is a <u>surface</u> everywhere tangent to the vorticity field.

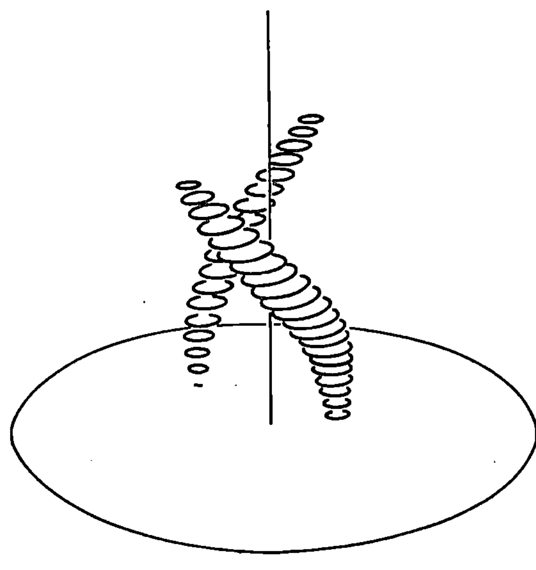

Figure 11.   Several vortex-lines
associated with the shear-layer
in the base flow (Fig. 7).

That the lines intersect the lower boundary at right angles is a
consequence of the lower boundary conditions, Eqs. (8) and (9) which
imply $\xi(r,0) = \eta(r,0) = 0$. The lines bend in the negative azimuthal
direction and slightly outward with increasing height until the <u>baffle</u>
layer is reached. At this level, ζ decreases rapidly to zero (because of
the upper conditions, Eq. (10)) and thus the lines bend over to the horizontal.

The hypothesis is that three-dimensional instabilities take the shape
of these vortex lines as will the ultimate finite amplitude motion.

There is some evidence that the mesocyclone also exhibits two vorticity
maxima and that these maxima are displaced clockwise with height. Figure 8
of [7] shows that two maxima in the vertical component of vorticity occur
and between z = .3 and z = 1.3 km the maxima appear to be displaced clockwise with height. Such is also the case in another Doppler-radar study of
a different storm [28]. See also the study by Klemp and Rotunno [32].

<u>Closing Remarks</u>

It is observed that tornadic thunderstorm updrafts undergo a basic
change in structure when a tornado forms. The updraft, before the tornado
forms, is broad and the center of circulation is near the updraft center.
However, when the tornado forms the updraft becomes contorted into a 'ring'
or 'horseshoe' shape surrounding downdraft. Tornadoes are observed to
occur at the edge of this ring between the updraft and the downdraft.

Using a simple three-dimensional model of a rotating updraft, we
found that

i)   the central downdraft develops in response to an intensifying low-
level circulation and,

ii)   as a result of three-dimensional instability, the axisymmetric
updraft with central downdraft develops smaller intense circulations situated between the updraft and the downdraft.

We have deliberately examined a simple flow in order to isolate several
fundamental mechanisms. The effects of a frictional lower boundary, density
variation and inflow asymmetry are all neglected. Further, because the
volume flow rate is held fixed (and thus is not dependent on the updraft
structure as it may in a thunderstorm), a steady state is achieved in our
numerical integrations. We study here, over a short time, a small portion
of the tornadic thunderstorm. The general problem is obviously much more
complicated.

Progress toward a better understanding of the tornado has been aided
by technological advances in radar meteorology and high-speed computers.
For example, a detailed comparison of velocity fields obtained from a
multi-Doppler radar interogation of a tornadic storm with those produced
by a three-dimensional numerical model of same are encouraging [28].

References

1. Ray, P.S., 1976: *J. Appl. Meteor.*, 15, 879-890.

2. Bluestein, H.B., 1980: *Bull. Amer. Meteor. Soc.*, 71, 560-567.

3. Lemon, L.R. and C.A. Doswell III, 1979: *Mon. Wea. Rev.*, 107, 1184-1197.

4. Burgess, D.W.: Personal communication.

5. Burgess, D.W., V.T. Wood and R.A. Brown, 1982: *Proc. 12th Conf. on Severe Local Storms*, San Antonio, 12-15 January.

6. Golden, J.H. and D. Purcell, 1978: *Mon. Wea. Rev.*, 106, 22-28, 422-424.

7. Brandes, E.A., 1978: *Mon. Wea. Rev.*, 106, 995-1011.

8. Lemon, L.R., D.W. Burgess, R.A. Brown, 1975: *Proc. 9th Conf. on Severe Local Storms*, Norman, 21-23 October.

9. Browning, K.A. and F. Ludlam, 1962: *Quart. J. Roy. Meteor. Soc.*, 88, 117-135.

10. Lewellen, W.S., 1971: NASA Rep. CR-177w, 219 pp.

11. Ward, N.B., 1972: *J. Atmos. Sci.*, 29, 1194-1204.

12. Davies-Jones, R.P., 1976: *Preprint, Symposium on Tornadoes*, Texas Tech University, 151-174.

13. Church, C.R., J.T. Snow, G.L. Baker and E.M. Agee, 1979: *J. Atmos. Sci.*, 36, 1755-1776.

14. Snow, J.T., 1982: *Proc. 12th Conf. on Severe Local Storms*, San Antonio, 12-15 January.

15. Rotunno, R. and D.K. Lilly, 1980: NRC Final Report NUREG/CR-1840, 51 pp.

16. Rotunno, R., 1977: *J. Atmos. Sci.*, 34, 1942-1956.

17. Rotunno, R., 1979: *J. Atmos. Sci.*, 36, 140-155.

18. Binnie, A.M. and G.A. Hookings, 1948: *Proc. Roy. Soc. London*, A194, 348-415.

19. Davies-Jones, R.P., 1973: *J. Atmos. Sci.*, 30, 1427-1436.

20. Baker, G.L. and C.R. Church, 1979: *J. Atmos. Sci.*, 36, 2413-2424.

21. Howard, L.N. and A.S. Gupta, 1962: *J. Fluid Mech.*, 14, 463-476.

22. Rayleigh, Lord (J.W. Strutt), 1880: *Proc. London Math Soc.*, 11, 57-70.

23. Rayleigh, Lord (J.W. Strutt), 1916: *Proc. Roy. Soc. London*, A93, 148-154.

24. Davies-Jones, R.P. and E. Kessler, 1974: *Weather and Climate Modification*, W.N. Hess, Ed., Wiley, 842 pp.

25. Snow, J.T., 1978: *J. Atmos. Sci.*, 35, 1660-1671.

26. Staley, D.O. and R.L. Gall, 1979: *J. Atmos. Sci.*, 36, 973-981.

27. Rotunno, R., 1978: *J. Fluid Mech.*, 87, 761-771.

28. Klemp, J.B., R.B. Wilhelmson and P.S. Ray, 1981: *J. Atmos. Sci.*, 38, 1558-1580.

29. Burgess, D.W., R.A. Brown, L.R. Lemon and C.R. Safford, 1979: *Proc. 10th Conf. on Severe Local Storms*, Omaha, 18-21 October.

30. Fujita, T., 1973: *Proc. 9th Conf. on Severe Local Storms*, Norman, 21-23 October.

31. Agee, E.M., J.T. Snow and P.R. Clare, 1976: *Mon. Wea. Rev.*, 104, 552-563.

32. Klemp, J.B. and R. Rotunno, 1981: *Proc. of the Joint IUTAM/IUGG Symposium on Intense Atmospheric Vortices*, Reading, England, 14-17 July.

by

P. G. Saffman
Applied Mathematics
California Institute of Technology
Pasadena, California U.S.A. 91125

Abstract

Results are presented for the steady motion of finite cored uniform vortices in an incompressible perfect fluid. The cases of single vortices, vortex pairs, single rows and the staggered double row are considered. The stability of the flows to two-dimensional disturbances is examined. In particular, it is shown that finite area stabilises the Karman vortex street for a finite band of aspect ratio values about the Karman value of 0.281.

1. Introduction

Since the classic work of Helmholtz and Kelvin over 100 years ago, here has been continuing interest in the properties of vortex arrays, and there are very many papers devoted to various examples of such motions. In particular, two-dimensional motion of incompressible, slightly viscous flow containing vorticity has been modelled many times as an assembly of point vortices. In this representation, it is assumed that the vorticity can be discretized into points (more strictly lines in the direction perpendicular to the plane of the flow) so that

$$\omega(x,y,t) = \sum_{j} \kappa_j \, \delta(r - r_j(t)) \; . \tag{1}$$

Here we employ the notation that κ_j denotes the circulation (strength) of the j-th vortex and $r_j = (x_j, y_j)$ is its position vector; δ is the two-dimensional Dirac delta function. The velocity field induced at (x,y) by this vorticity field has components u, v given by

$$u = \frac{-1}{2\pi} \sum_{j} \kappa_j \, \frac{y - y_j}{|r - r_j|^2} \; , \quad v = \frac{1}{2\pi} \sum_{j} \kappa_j \, \frac{x - x_j}{|r - r_j|^2} \; . \tag{2}$$

The expression (2) is for unbounded fluid. In the presence of walls or other bounding surfaces, the flow due to image vortices or some other appropriate irrotational velocity field has to be added to (2). The evolution of the vorticity distribution is then given by the Helmholtz laws, from which it follows that the strengths κ_i are constant and the velocities $\dot{r}_i$ of the vortices are given by the velocity field (2) evaluated at $r = r_i$ with the infinite self-induced component subtracted. The calculation of the motion is in this way reduced to the solution of a set (possibly infinite) of ordinary differential equations.

There are many elegant analytical solutions of such systems corresponding to different flow geometries and problems. One of the better known and thoroughly studied cases is the double staggered periodic row of equal and opposite vortices known as the Karman vortex street which models the wake of a bluff body in a uniform stream for Reynolds numbers around 100; see [1] Ch. XIII. However, the assumption that the vorticity is

collected into a set of delta functions raises doubts about the accuracy of
the approach or its ability to model the continuous vorticity distributions
of a real flow. For instance, it was shown by Karman that this street is
unstable to infinitesimal two-dimensional disturbances unless the aspect
ratio h/ℓ has the value 0.281, where h is the distance between the rows and
ℓ the separation between neighboring vortices in the same row, and later
shown by Kochin that even this case is unstable to finite amplitude
two-dimensional disturbances; but observations indicate values in the range
0.2 - 0.4 and at most a weak instability is seen [2].

Recently, a number of workers have approached directly the problems of
motion of vortex arrays and vortices with the vortices given a finite core,
but still retaining for the most part the simplifications of two-dimensional,
inviscid, incompressible flow. The present article will review briefly some
of the known results for 'simple' cases, namely the isolated vortex and the
co-rotating vortex pair, and then describe new results and the status of
current work for vortex streets of finite cored vortices. Some rather
interesting results have been obtained in these studies, but a number of
open questions or controversies remain and the field is active. One point
should perhaps be emphasized. The use of finite cored vortices must be an
improvement over the use of point vortices for the modelling of vortex
motion, but to date it has been the practice to take the vorticity
distribution as either uniform inside the core or concentrated entirely on
the surface so the core is also irrotational or hollow but separated by a
vortex sheet from the external irrotational flow. Real vortices seem to
have significant variations of vorticity inside them, and the modelling
still therefore seems to leave something to be desired. In principle a more
realistic vorticity distribution could be studied by dividing up each vortex
into several regions each containing uniform but different vorticity and the
methods to be described here should work just as well; but the amount of
computing will be substantially increased and for this reason the piecewise
uniform vortex has not yet been employed. It is hoped, however, that the
increased availability of computing resources will allow this approach to be
employed in the near future.

We wish to mention in passing that the computer allows the motion of
finite cored vortices to be studied by replacing them with a very large
number of point vortices whose motion is then followed either by the cloud
in cell method, e.g. [3], or by integrating the set of ordinary differential
equations, e.g. [4]. The latter method makes even more critical the
question of the relation between the solutions of the Euler equations for
piecewise continuous vorticity, or singular vorticity on a line corresponding
to a vortex sheet, and integrals of the ordinary differential equations
derived either from (2) or modifications that remove the singularity as
$\underset{\sim}{r} \to \underset{\sim}{r_i}$. A further but crucial complication is that only approximate
integrals are evaluated, and the relation between the exact and numerical
integrals of the ordinary differential equations needs to be established.
Attempts, e.g. [5], at some of these difficult questions have been made, but
it is not clear whether the cases of fluid dynamical interest satisfy the
conditions for the given theorems to apply.

2. The Isolated Vortex

Consider the case of a single uniform vortex in an unbounded flow. The
basic solution is the Rankine vortex, with vorticity of magnitude ω_0 inside
a circle $r = R$ giving a velocity distribution in polar coordinates

$$u_\theta = \frac{\omega_0 r}{2} \; , \quad r < R \; , \quad u_\theta = \frac{\omega_0 R^2}{2r} \; , \quad r > R \; . \tag{3}$$

As is well known, this is not a unique solution for a uniform vortex of given area, unless circular symmetry is imposed. As shown by Kirchhoff, see [6] p.232, there is an elliptical vortex solution, in which the vorticity is confined inside an ellipse of semi-axes a, b which rotates with angular velocity $\Omega = \omega_0\, ab/(a+b)^2$. It was recently demonstrated by Deem and Zabusky [7], that there are infinitely many other solutions with m-fold rotational symmetry, where the ellipse corresponds to m = 2, equilateral triangular shapes to m = 3, square like shapes to m = 4, and so on. These solutions are bifurcations of the circular shape associated with the normal modes

$$r = R(1 + \varepsilon \cos(m\theta - \Omega_m t))$$

where $\Omega_m = \tfrac{1}{2}\omega_0(m-1)/m$. The angular velocity of the m-th state is believed to lie in a band

$$\Omega_m^c < \Omega \le \Omega_m \tag{4}$$

with the vortex becoming more deformed and more like an m-sided, curvilinear polygon as Ω decreases from Ω_m and approaches a critical value Ω_m^c. Rigorous proofs of these beliefs do not appear, however, to have been given and apart from the m = 2 case, where the solution exists in closed form, the evidence is numerical; see [8] and [9]. It is claimed on the basis of the calculations that $\Omega_m^c = \Omega_{m-1}$ and further that shapes are stable to infinitesimal two-dimensional perturbations if $\Omega > (3\Omega_m + \Omega_{m-1})/4$. For m = 1, the claims are consistent with Kirchhoff's solution and with Love's result that the Kirchhoff elliptical vortex is stable for a/b < 3. (If these results are true for m > 1, their simple form suggests that they should be capable of analytical proof.)

The Kirchhoff vortex was generalised by Moore and Saffman [10] to describe elliptical vortices in uniform straining fields of infinite extent. There are reasons based on the dispersion relation for infinitesimal perturbations to believe strongly that the elliptical solutions bifurcate into non-elliptical solutions if the deformations are sufficiently large, but these have not yet been calculated.

It should be noted that Kida [11] has extended the Moore and Saffman elliptical solutions to unsteady motion. If the principal axes are a(t), b(t), and $\theta(t)$ is the angle between the axis of length a(t) and the direction of maximum extension, then for the case of time independent irrotational strain with extension rate ε,

$$\left.\begin{aligned}
\dot{a} &= \varepsilon\, a \cos 2\theta, \qquad \dot{b} = \varepsilon\, b \cos 2\theta \\[2mm]
\dot{\theta} &= -\,\varepsilon\, \frac{a^2 + b^2}{a^2 - b^2}\sin 2\theta + \frac{ab\omega_0}{(a+b)^2}\ .
\end{aligned}\right\} \tag{5}$$

These equations allow the possibility of improving applications of the so-called elliptical vortex model [12] to unsteady motion.

3. Methods

It is appropriate to outline the numerical methods that are employed, as these apply to all cases of finite cored uniform vortices with suitable modifications. One method, introduced by Deem and Zabusky [7], following the water bag method of plasma physics, uses an integration by parts of the Green's function to give the complex velocity field u + iv as a function of $z = x + iy$ due to a single uniform vortex as a contour integral around the boundary:

$$u + iv = - \frac{\omega_0}{2\pi} \int \log|z - Z(s)| \, dZ(s) \tag{6}$$

where $Z(s)$ is the parametric equation of the boundary and s is arclength, say. The velocities of more vortices simply add linearly and a piecewise uniform vortex can be regarded as the superposition of a finite number of uniform vortices. Let us denote by U, V the values of u, v given by the sum of terms like (6) evaluated on the boundary of the vortex. Then the condition that the vortex moves with the fluid, or that the boundary is a streamline when the flow is steady, gives the governing equation

$$\text{Im} \left[\left\{ \frac{\partial Z}{\partial t} - (U + iV) \right\} \frac{dZ^*}{ds} \right] = 0 \ . \tag{7}$$

The majority of the numerical work to date solves this equation numerically by iteration or Newton's method for some finite representation or discretization of the boundary. Burbea, see [9], employs a method in which the basic unknowns are the coefficients in the Taylor series expansion of the analytic function which maps the exterior of the vortex into the exterior of the unit circle.

An alternative approach is to expand the solution as a power series in the area A of the vortices, taking as the basic solution small circular vortices. The algebra quickly becomes difficult and extremely laborious, but Kida [13] has employed this method to study the stability of the Karman vortex street (although his results do not completely agree with those given in §6 below). Currently, work is in progress (Jimenez, private communication) to see if the algebra of the conformal map representation can be automated for computer calculation to enable many terms of the series expansion to be found.

The techniques of Kirchhoff-Helmholtz free streamline theory can be applied to hollow or stagnant cored vortices, e.g. [14].

Three methods have been applied to the study of stability to two-dimensional disturbances:

Method I is the spectral approach, which calculates the eigenvalues and eigenfunctions for infinitesimal disturbances to steady (in some reference frame) states. The method has the advantage of being definitive, but suffers from some disadvantages. Firstly, it tends to be expensive and the undisturbed state may need to be known extremely accurately (to 0.1% in some cases) for the eigenvalues to be found reliably. Secondly, it does not enable one to predict what the eventual end state may be if the calculation predicts instability. Thirdly, since there is no dissipation, stability only means neutral stability and the behavior of finite amplitude disturbances is unknown.

Method II is global and powerful when it works, but unfortunately that is not too often. It employs Kelvin's variational principle, according to which steady states are those for which the kinetic energy of the flow induced by the vorticity field is an extremum for isovorticial variations [15]. If it can be shown that the extremum is a maximum or minimum, then stability follows.

Method III is to perturb the undisturbed state and calculate the evolution in time of the vortices using one of the time-dependent numerical codes for finite vortices, e.g. cloud in cell [3] or contour dynamics [8]. This is also expensive and suffers from the disadvantage that only a few initial states can be considered, so instabilities may be missed, or predicted incorrectly when the initial disturbance is too large. Also, the problem of distinguishing between numerical and hydrodynamical instability can be difficult.

In short, each approach has its advantages and disadvantages, but each will give useful information for stability to two-dimensional disturbances. The two-dimensional stability of the isolated vortex has been studied by both methods I [9] and III [8].

A study of stability to three-dimensional disturbances is just beginning in a serious way. When the vortices are small and the axial wavelength is long, the Biot-Savert law for vortex filaments can be employed. This was done over 50 years ago by Schlayer and Rosenhead for the Karman vortex street and later by Gupta, Crow and Jimenez for vortex pairs. Recently, the calculations have been repeated for the street and other arrays [16]. The case of axial wavelength comparable with the core diameter has been studied by Tsai and Widnall [17] and Moore and Saffman [18] for the isolated vortex weakly distorted by a straining field, and very recently Pierrehumbert and Widnall [19] have investigated the stability of the Stuart vortex row. Other work is in progress, and more information on three-dimensional stability should become available before too long. It is probable that the existence of new steady three-dimensional vortex motions (other than the helical vortex and vortex ring) will be uncovered as bifurcations from two-dimensional states.

4. The Vortex Pair

The motion of two finite cored vortices has been studied by several authors. For the case of the translating pair of equal and opposite uniform vortices, a solution was given by Deem and Zabusky [7] and an extensive investigation by Pierrehumbert [20]. The rotating motion of two equal uniform vortices was calculated by Saffman and Szeto [21]. For results with non-equal vortices, see [8].

There is an open question here concerning the translating pair and its limit as the vortices increase in size or get close together. Pierrehumbert [20] claims that there is a limit for $A/\ell^2 = 59.2$, where ℓ is the distance between the centroids, with the vortices touching along a common axis. Saffman and Szeto [21] raise the possibility that this solution is isolated and that there exist solutions for arbitrarily large A/ℓ^2, as for hollow or stagnant cored vortices [22]. Numerical methods have so far proved inadequate to resolve this controversy.

The results for the co-rotating pair show that there is a unique solution for $A/\ell^2 < 4.93$. For $4.93 < A/\ell^2 < 5.00$, there are two solutions, with limit point behavior at the upper limit. In other words, the separation distance between the ends of the vortices is not a monotonic function of A/ℓ^2, but decreases from infinity to zero as A/ℓ^2 increases from zero, reaches a maximum of 5.00, and then decreases to 4.93 as the vortices touch.

The stability of the translating pair has been studied by method II [21]. The energy is not an absolute maximum or minimum, but it may, however, be a relative maximum or minimum and hence stable, if there exist non-symmetrical steady solutions. None of these have been found, but they may exist, and so the question of stability remains open. If, however, only symmetrical disturbances are allowed, as is appropriate if the pair was a vortex and its image in a wall, then method II predicts that the pair is stable. For the co-rotating pair, method II predicts that the solution is

stable except that the solution with smaller separation is unstable when the
solution is not unique. This is in agreement with calculations using method
III [3].

5. The Single Row

The array consisting of an infinite number of equal uniform vortices
centered at equally spaced intervals on a straight line was investigated
independently by Saffman and Szeto [23], who also studied stability by
method II, and Pierrehumbert and Widnall [24]. The hollow or stagnant array
was described in [14], where the stability was also investigated by method I
(see also [23]) and the predictions of method II for this case confirmed.
The results are interesting and have some relevance to the behavior of the
coherent structures seen in the turbulent mixing layer as described, for
example, in [25].

The array is determined by the value of A/ℓ^2, where A is the area of
each vortex and ℓ the separation distance. The strength of the vortices
just scales the magnitude of the velocities. For $A/\ell^2 < 0.203$, the solution
is unique and by method II stable to superharmonic disturbances, i.e.
perturbations in which all the vortices are distorted in an identical
manner. In the range, $0.203 < A/\ell^2 < 0.238$ there are two solutions with
limit point behavior at the upper limit. This is a change of stability with
respect to superharmonic disturbances, and the unstable vortices are said to
be unstable to the tearing instability as it is expected that the
disturbance will cause the vortices to disintegrate when it reaches finite
amplitude.

The array is always unstable to subharmonic disturbances with
wavelength 2ℓ. This is the so-called pairing instability. It can be
calculated exactly for A = 0 and there is no reason to believe that the
results are significantly altered by finite area effects, based on the
evidence of unsteady simulations and the finite area results for hollow
vortices [21].

The question of relevance to the understanding of the turbulent mixing
layer is to determine which of the two instability mechanisms is more
relevant to the evolution of the coherent structures. This has been
discussed in [26], where it is argued on the basis of energy considerations
that the pairing takes place when the array is young and tearing when its
old, age being measured by the amount of energy that has been dissipated.

It would be of interest to study the growth rate of subharmonic
disturbances of wavelength other than 2ℓ. The evidence is that 2ℓ is the
wavelength of the most unstable disturbance when A is small, but this may
not be the case when A is finite. The occasional interaction of three
vortices suggests that the disturbance wavelength 3ℓ may have comparable
gorwth rate. It would also be valuable to investigate the properties of
piecewise uniform vortices for the purpose of studying the possible effects
of non-uniform vorticity.

Another calculation worth doing is the structure and stability of a
curved row. This could be modelled by investigating the array formed of
finite cored vortices placed at the vertices of a polygon. Experiment has
shown that curved mixing layers, concave to the slower moving fluid, grow
less rapidly. It is known that a curved row of point vortices is stabilised
when the radius of curvature R is sufficiently small; according to Thomson's
calculation (see [27]), the condition is $\ell > 2\pi R/7$. The effect of finite
area on this result is of interest.

6. The Karman Vortex Street

We now present results on the properties of the double staggered array. The geometry is sketched in figure 1. It is clear that symmetrical solutions will exist when the offset stagger d is zero or equal to $\frac{1}{2}\ell$. We shall only consider the latter case, as the indications are that this is the only possible stable state. We shall also suppose the vortices are all of equal area and the strengths have the same magnitude but opposite signs in the two rows.

A frame of reference is chosen with respect to which the vortices are stationary. Since the street moves relative to the fluid at infinity, the velocity at infinity then has this value U_s, say. The sum of (6) over the vortices gives the expression for the velocity

$$u + iv = \frac{\Gamma}{2\pi A} \oint \log\left|\sin\frac{\pi}{\ell}(z-Z)\sin\frac{\pi}{\ell}\left(z-Z-\frac{\ell}{2}-ih\right)\right| dZ + U_s, \qquad (8)$$

where the integral is around one of the vortices in the upper row and it is assumed that the vortices in the lower row are reflections of those above. The basic equation for the steady states is

$$\text{Im}\left[\frac{dZ^*}{ds}(U+iV)\right] = 0 . \qquad (9)$$

The solution is a function of two parameters; $\alpha = A/\ell^2$ and $\kappa = h/\ell$. Again the magnitude of Γ just scales the velocity.

This equation can be solved approximately by taking the origin at the center of the vortex and writing $Z = R(\theta)e^{i\theta}$. The approximation $Z = \sum_0^N a_n \cos n\theta$ is substituted and (9) is satisfied at N points. The resulting equations together with the condition that the centroid is at the origin and that the area is given provide N+2 equations for the N+2 unknowns $a_0, a_1, \ldots a_N$ and U_s. These are solved by iteration or Newton's method. Continuation in α was employed to give first guesses; for α small the first guess assumed circular shapes and the value of U_s for point vortices. Accuracy of the solutions was checked by requiring that the results are independent of the number of Fourier coefficients retained.

Some typical shapes are shown in figure 2 and results for $\hat{U}_s = \ell U_s/\Gamma$ in figure 3. The solution is in general unique, but for relatively large κ and α there is a small region where two solutions exist. It appears that there is an upper bound on the value of α, depending on κ, for the solutions to exist. The limit is characterized by vortices touching; for larger κ it is the vortices in the same row, for smaller κ it is those in different rows, which touch. For further details, see [28].

There are a number of open questions regarding double rows. Do non-symmetrical solutions or arrays with stagger d not equal to zero or $\frac{1}{2}\ell$ and velocity parallel to the row exist? None has yet been found, but neither has an exhaustive search been made. The double row of unequal vortices models the wave of a bluff body in a shear flow, since it is a combination of mixing layer and wake. This has been modelled in a time dependent calculation using clouds of point vortices in [29]. Do steady solutions exist? It is easy to see that there is no solution if the areas of the vortices are zero, but perturbation methods and the elliptical vortex model indicate that finite area solutions exist provided the areas are greater then some minimum value proportional to the difference in vortex strengths.

We turn now to the stability of the Karman vortex street to two-dimensional disturbances. Let $\psi(x,y)$ denote the stream function of the steady motion in street fixed coordinates for some configuration of the array. The question is as follows: if the stream function ψ is perturbed about Ψ and has the form

$$\psi(x,y,t) = \Psi(x,y) + \epsilon \psi'(x,y,t) , \qquad (10)$$

does ψ' grow as t increases or stay bounded? (In the absence of dissipation, it is not possible for the energy of the disturbance to vanish, although ψ' could vanish pointwise if the spectrum were continuous.)

The problem was studied by Christiansen and Zabusky [30] using the initial value unsteady calculation of method III. They found that at the Karman value of $\kappa = 0.281$, disturbances of wavelength 2ℓ appear to be stabilised by finite area. We now present results of the spectral approach method I, obtained by finding ψ' when ϵ is infinitesimal, which confirms these calculations and gives further information, although some open questions still remain.

Since $\Psi(x,y)$ is independent of t and has period ℓ in x, it follows that the normal mode ψ' has the form

$$\psi'(x,y,t) = e^{i2\pi px/\ell} \cdot e^{\sigma t} \sum_{-\infty}^{\infty} \phi_n(y) e^{i2\pi nx/\ell} , \qquad (11)$$

where σ is the eigenvalue and p is an arbitrary real number. The functions ϕ_n constitute the eigenvector. The requirement that the linear equation for ψ' has a non-trivial solution with the velocity everywhere continuous and the vorticity just convected by an incompressible displacement determines σ as a function of p, α, κ and a mode label M, since we expect an enumerable infinity of eigenvalues for each configuration, i.e.

$$\sigma = \frac{\Gamma}{\ell^2} \hat{\sigma}(p,\alpha,\kappa;M) . \qquad (12)$$

Note that there is no loss of generality in supposing that $0 \leq p < 1$, as changing p by an integer just relabels the components ϕ_n of the eigenvector.

If p is zero, the perturbation stream function is also of period ℓ and all vortices are perturbed in the same way. Such disturbances are called superharmonic. For small α, these oscillations are like the vibrations of an isolated vortex. We expect for these that σ is pure imaginary and proportional to Γ/A.

If p is not zero, the disturbance has components with many wavelengths, but in particular there is a component with wavelength ℓ/p. These disturbances will be called subharmonic.

Note that if σ is an eigenvalue for a particular value of p, so is $-\sigma$ since the Euler equations are time reversible, and so also is σ^* for the value $-p$. But $-p$ and $1-p$ are equivalent (apart from a relabelling of the eigenvector), so

$$\hat{\sigma}(p,\alpha,\kappa;*) = \hat{\sigma}^*(1-p,\alpha,\kappa;*) \qquad (13)$$

for some value of the mode label.

If $\mathrm{Re}(\sigma) \neq 0$ for any eigenvalue, then the state is unstable. Neutral stability requires $\mathrm{Re}(\sigma) = 0$ for all eigenvalues, and implies stability to infinitesimal disturbances in the sense that they stay bounded. Whether

this means stability to small but non-infinitesimal disturbances is a
difficult question which will not be considered further here.

For zero area, $\alpha = 0$, only the subharmonic disturbances exist and
there is only one mode, the zero mode, say. In this case $\hat{\sigma}$ can be
calculated explicitly ([6], §156) and the stability boundary separating
$\text{Re}(\sigma) = 0$ and $\text{Re}(\sigma) \neq 0$ can be calculated. This is shown in figure 4 in
the p-κ plane. Note the symmetry about $p = 0.5$.

The case $\alpha = 0$ is singular and the question is the behavior when $\alpha \neq 0$.
It is expected that only the zero mode similar to the $\alpha = 0$ subharmonic
disturbance is important; the other modes correspond to rapid stable
oscillation of the vortex boundary and can presumably be neglected. We are
therefore interested in the behavior of the stability boundary when $\alpha \neq 0$.
From our general qualitative considerations based on symmetry about $p = 0.5$
and the non-singular behavior for $\alpha > 0$, it follows that the degenerate.
saddle point behavior of figure 4 will separate into one of the two
possibilities shown by dashed lines. If case 1 is the situation, there will
be stability to infinitesimal disturbances for a finite range of κ in the
vicinity of $\kappa = \kappa_c$. If case 2 obtains, then finite size makes the array
unstable for all κ. We wish to emphasize that the symmetry about $p = 0.5$
holds for all α, so cases 1 and 2 are the only possibilities and it is
sufficient for the purpose of establishing stability of the array to
subharmonic disturbances to limit attention to the case $p = 0.5$ or doubling
the wavelength. Moreover, $d(\text{Re}(\sigma))/dp = 0$ when $p = 0.5$, and hence at least
for small area the most unstable disturbance has $p = 0.5$.

This case was analysed by introducing the same perturbation to every
other vortex in each row, there being then four different shapes in the
perturbation and correspondingly four rows each with wavelength 2ℓ. The
radius of each vortex was perturbed by an expression

$$\epsilon \sum_0^\infty (a_n^{(m)} \cos n\,\theta + b_n^{(m)} \sin n\,\theta), \text{ where } m = 1, 2, 3, 4. \text{ The series was}$$

truncated to N terms, and evaluation of the velocity field accomplished by
summing and integrating over the 4 rows. The unsteady equation (7) was then
satisfied at $2N+1$ points on each of the four vortices, giving $8N+4$ linear
equations in the $8N+4$ unknown Fourier coefficients. This is a generalized
eigenvalue problem of the form $\underline{A}\underline{w} = \sigma\,\underline{B}\underline{w}$, where $\underline{A}$ and $\underline{B}$ are $8N+4$ matrices
and w contains the $8N+4$ coefficients.

An IMSL library routine was used to calculate the value of σ. Values
of N from 10 to 25 were employed. Eigenvalues were regarded as physically
sensible only if they were independent of N and could be associated with
the known modes for small area. The zero mode was easily recognizable.
The high frequency oscillations behaved like $\sigma = \tfrac{1}{2} i\,\Gamma(m-1)/2\pi A$ for
$m = 2, 3, 4, \ldots$.

The results of the calculation are summarized in figure 5. Outside
of the indicated region in the (κ,α)-plane, there are modes with positive
growth rate and the street is unstable. Inside the region, the $p = 0.5$
mode is stable, and our argument is that this implies that <u>all</u> subharmonic
modes are stable. For small area, stability occurs for

$$\kappa_c - .58\alpha^2 < \kappa < \kappa_c + 1.64\alpha^2 . \tag{14}$$

Thus finite area stabilizes the street.

The present method is limited in principle to modes with p rational,
and in practice p has to be the reciprocal of a small integer to keep the

amount of computing reasonable. Kida's [13] method of expanding in powers
of A allows the subharmonic stability to be calculated for arbitrary p, but
unfortunately his results do not show the symmetry about p = 0.5 which is to
be expected on the basis of the above arguments. His results are therefore
questionable.

Acknowledgement

The work described here was supported by the Department of Energy
(Office of Basic Energy Sciences). Control Data Corporation are thanked for
the granting of time on the Cyber 203 at the CDC Service Center, Arden
Hills, without which the larger calculations would not have been possible.
The results on the Karman vortex street are joint work carried out with
Dr. J. C. Schatzman.

References

[1]   S. Goldstein 1965 Modern Developments in Fluid Mechanics. Dover.

[2]   T. Matsui and M. Okude 1982 Vortex pairing in a Karman vortex street.
       (To appear).

[3]   J.P. Christiansen 1973 Numerical simulation of hydrodynamics by the
       method of point vortices. J. Comp. Phys., 13, 363-379.

[4]   K. Kuwahara and H. Takami 1973 Numerical studies of two-dimensional
       vortex motion by a system of point vortices. J. Phys. Soc. Japan 34,
       247-253.

[5]   O. Hald and V.M. Del Prete 1978 Convergence of vortex methods for
       Euler's equations. Math. Comp., 32, 791-809.

[6]   H. Lamb 1932 Hydrodynamics. Sixth Ed. Cambridge Univ. Press.

[7]   G.S. Deem and N.J. Zabusky 1978 Vortex waves; stationary V states,
       interactions, recurrence and breaking. Phys. Rev. Lett., 40, 859-862.

[8]   N.J. Zabusky 1981 Recent developments in contour dynamics for the
       Euler equations. Proc. IV Int. Conf. on Collective Phenomena. New
       York Academy of Sciences.

[9]   J. Burbea and M. Landau 1982 The Kelvin waves in vortex dynamics and
       their stability. J. Comp. Phys. (to appear).

[10]   D.W. Moore and P.G. Saffman 1971 Structure of a line vortex in an
       imposed strain. Aircraft Wake Turbulence (Eds. J.H. Olsen,
       A. Goldburg and M. Rogers) p.339-354. Plenum Press.

[11]   S. Kida 1981 Motion of an elliptic vortex in a uniform shear flow.
       J. Phys. Soc. Japan 50, 3517-3520.

[12]   P.G. Saffman 1979 The approach of a vortex pair to a plane surface in
       inviscid fluid. J. Fluid Mech., 92, 497-503.

[13]   S. Kida 1982 Stabilizing effects of finite core on Karman vortex
       street. (Unpublished manuscript.)

[14] G.R. Baker, P.G. Saffman and J.S. Sheffield 1976 Structure of a linear array of hollow vortices of finite cross section. J. Fluid Mech., 74, 469-476.

[15] V.I. Arnold 1980 Mathematical Methods of Classical Mechanics. Springer Verlag.

[16] A.C. Robinson and P.G. Saffman 1982 Three-dimensional stability of vortex arrays. (Unpublished manuscript.)

[17] C.-Y. Tsai and S.E. Widnall 1976 The stability of short waves on a straight vortex filament in a weak externally imposed strain field. J. Fluid Mech., 73, 721-733.

[18] D.W. Moore and P.G. Saffman 1975 The instability of a straight vortex filament in a strain field. Proc. Roy. Soc. A 346, 413-425.

[19] R.T. Pierrehumbert and S.E. Widnall 1982 The two- and three-dimensional instabilities of a spatially periodic shear layer. J. Fluid Mech., 114, 59-82.

[20] R.T. Pierrehumbert 1980 A family of steady translating vortex pairs with distributed vorticity. J. Fluid Mech., 99, 129-144.

[21] P.G. Saffman and R. Szeto 1980 Equilibrium shapes of a pair of equal uniform vortices. Physics of Fluids 23, 2339-2342.

[22] H.C. Pocklington 1895 The configuration of a pair of equal and opposite hollow straight vortices, of finite cross-section, moving steadily through fluid. Proc. Camb. Phil. Soc., 8, 178-187.

[23] P.G. Saffman and R. Szeto 1981 Structure of a linear array of uniform vortices. Stud. App. Math., 65, 223-248.

[24] R.T. Pierrehumbert and S.E. Widnall 1981 The structure of organized vortices in a free shear layer. J. Fluid Mech., 102, 301-313.

[25] A. Roshko 1976 Structure of turbulent shear flows: a new look. AIAA Journal 14, 1349-1357.

[26] P.G. Saffman 1981 Vortex interactions and coherent structures. Transition and Turbulence (Ed. R.E. Meyer) p.149-166. Academic Press.

[27] T.H. Havelock 1931 The stability of motion of rectilinear vortices in ring formation. Phil. Mag. (7) 11, 617-633.

[28] P.G. Saffman and J.C. Schatzman 1982 Stability of a vortex street of finite vortices. J. Fluid Mech., 117, 171-185.

[29] D.R. Boldman, P.F. Brinich and M.E. Goldstein 1976 Vortex shedding from a blunt trailing edge with equal and unequal external mean velocities. J. Fluid Mech., 75, 721-735.

[30] J.P. Christiansen and N.J. Zabusky 1973 Instability, coalescence and fission of finite-area vortex structures. J. Fluid Mech., 61, 219-243.

<u>Figure captions</u>

Figure 1.   Sketch of the infinite double row of uniform vortices.

Figure 2.   Exact shapes of vortices in a Karman vortex street. Arrows
             denote velocity vectors. (a) $\kappa = 0.3$, $\alpha = 0.254$. (b) $\kappa = 0.1$,
             $\alpha = 0.080$.

Figure 3.   Velocity of the street as a function of vortex size for various
             κ. Note that for κ and α large, there are two solutions.
             Dotted line denotes values assuming circular vortices.

Figure 4.   The stability boundary in the periodicity spacing-ratio
             (p, κ)-plane for point-vortex configurations. The growth rates
             are symmetric about $p = 0.5$ which gives also the maximum rate of
             growth for all κ. Curves 1 and 2 show qualitatively the two
             possible perturbed boundaries for small area. The calculations
             indicate that curve 1 is the true result.

Figure 5.   A plot of the area spacing-ratio (α, κ)-plane. Curve 1 denotes
             the maximum area for a given spacing ratio. The segment
             corresponding to smaller κ should be regarded as a lower bound
             for the maximum area because slow convergence in this region
             made it difficult to find the precise α at which vortices
             touched. Above curve 2, there are two solutions for a given
             pair (α, κ). The enclosed central region has neutral linear
             stability; configurations outside this region are linearly
             unstable. Curve 4 is an estimate of the boundary of the region
             in which the symmetrical $(d = 0)$ array can exist. Curve 3 gives
             the values for which the energy of the staggered array equals
             that of the symmetric array of twice the streamwise spacing,
             twice the area and same distance between the rows. The former
             is greater above the curve. It has been speculated
             (but probably incorrectly) that the symmetric state might be
             preferred between curves 3 and 4. x and o denote stable and
             unstable states according to [30].

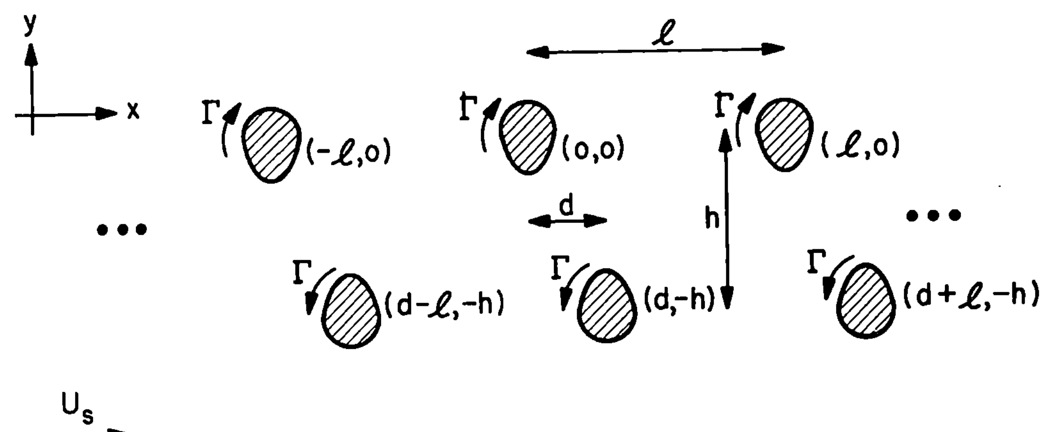

Figure 1

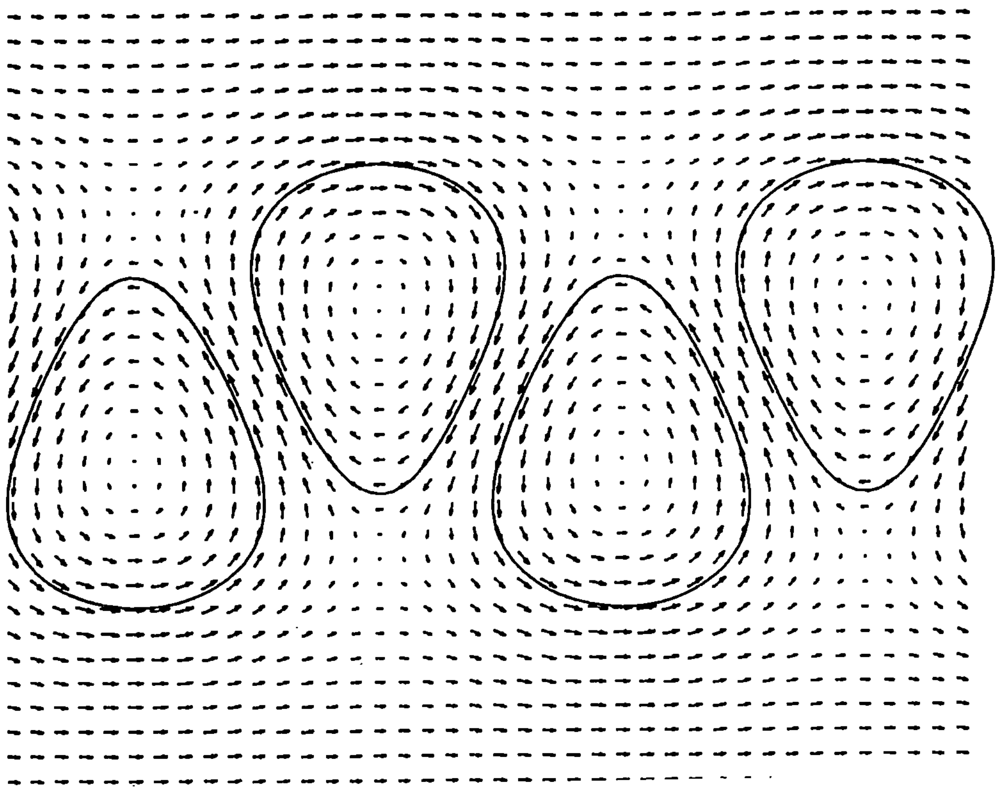

Figure 2a

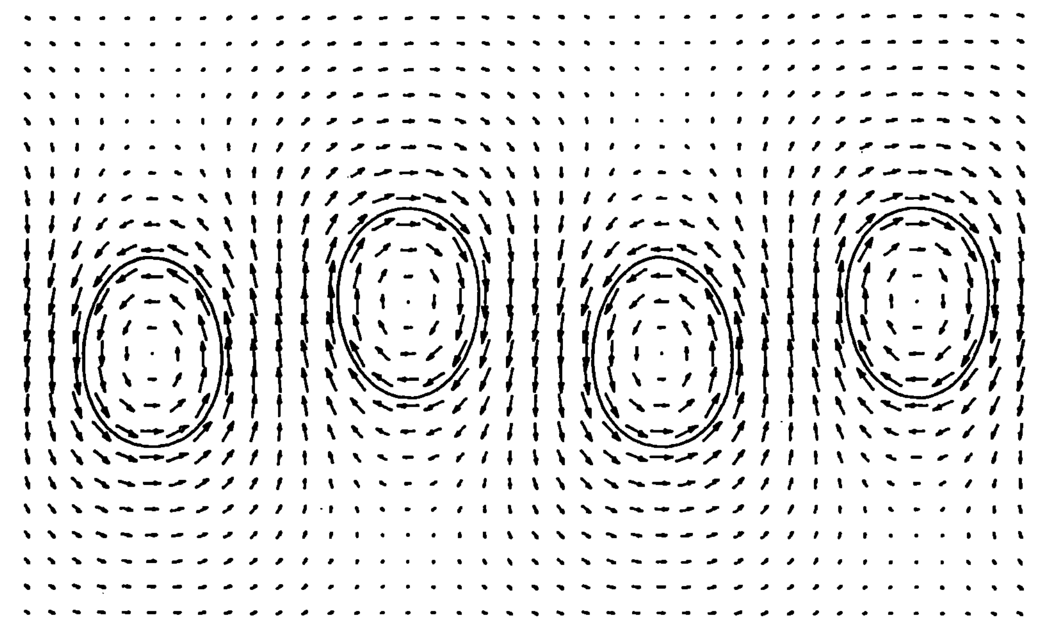

Figure 2b

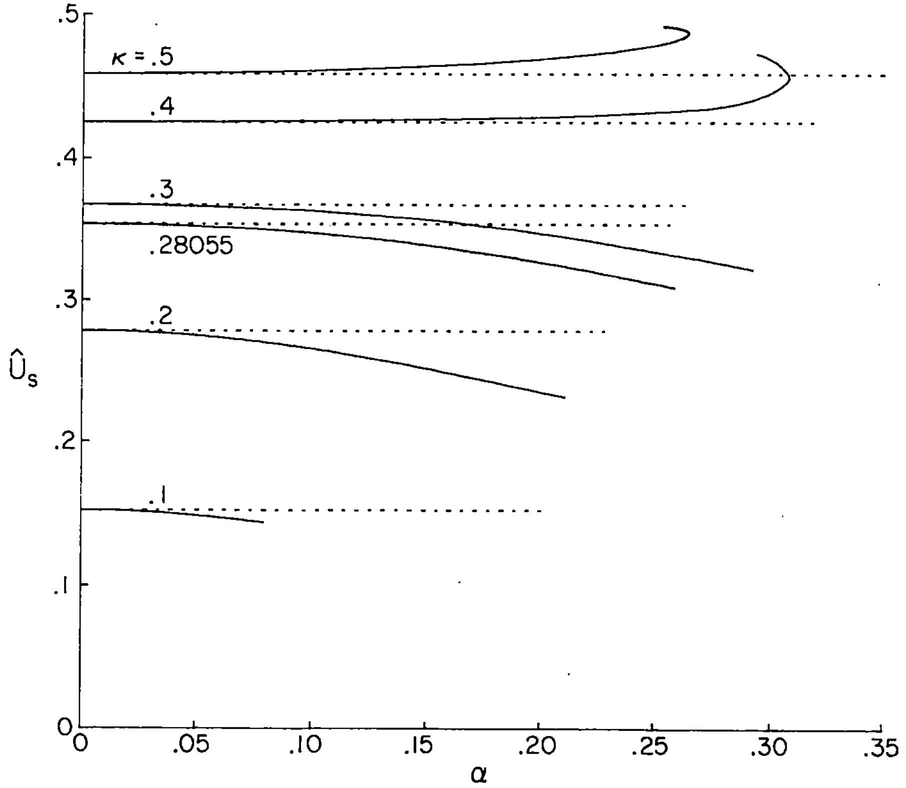

Figure 3

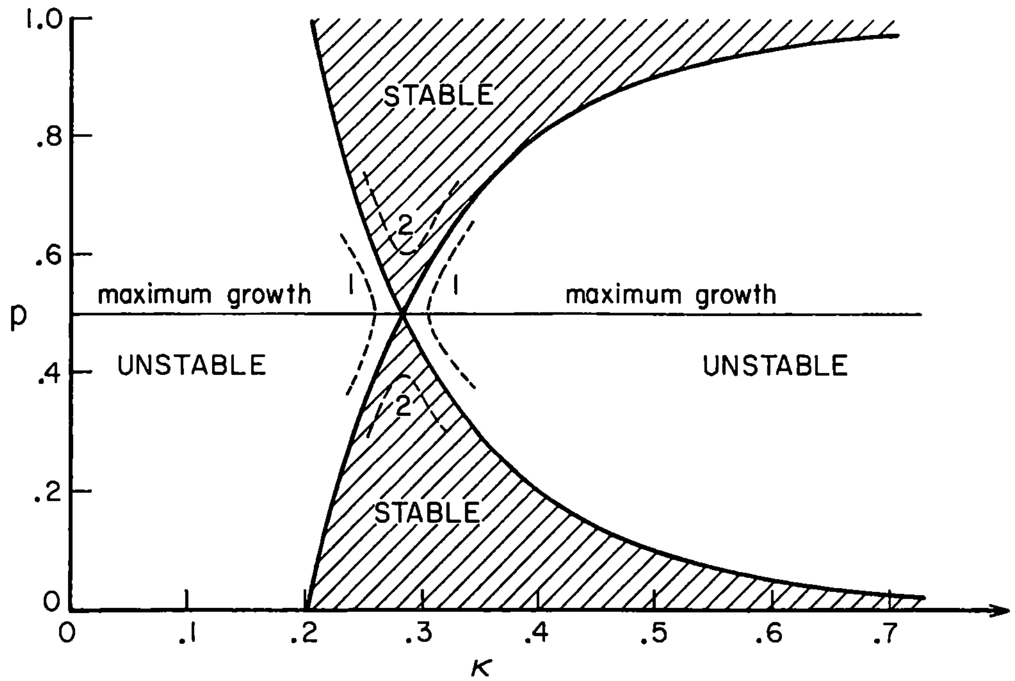

Figure 4

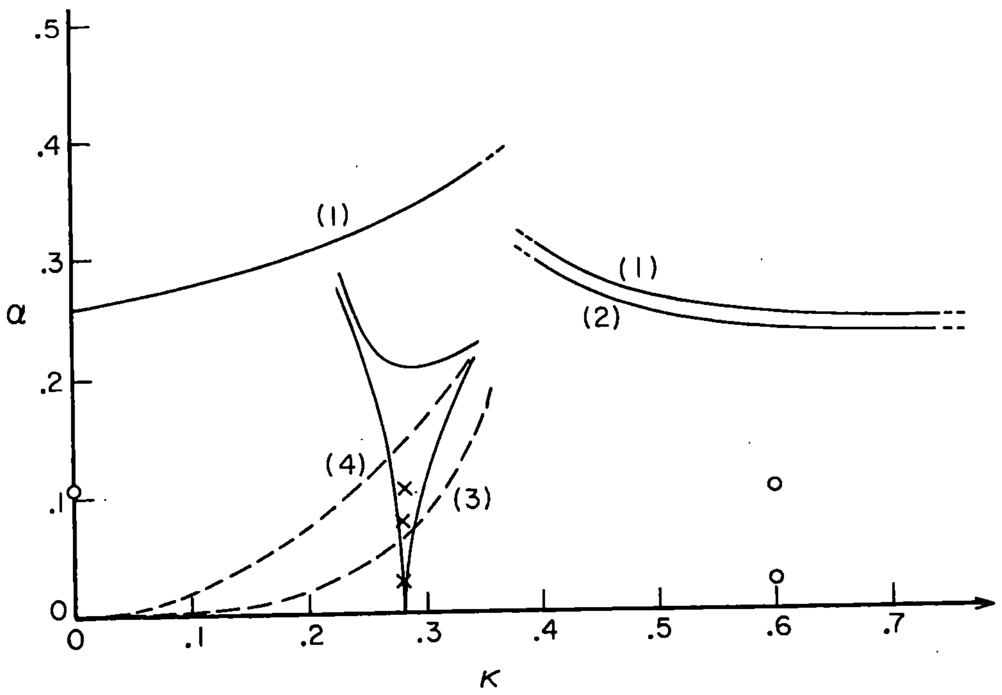

Figure 5

THE REPRESENTATION OF PLANAR SEPARATED FLOW BY

REGIONS OF UNIFORM VORTICITY

by

J. H. B. Smith

Aerodynamics Department, Royal Aircraft Establishment,
Farnborough, Hants, U.K.

Summary

After some introductory remarks about the role of inviscid models in
the representation of separated flow in two and three dimensions, an
examination is made of the local behaviour of an inviscid model of planar
separated flow. It is supposed that a body, together with a closed wake, is
surrounded by an irrotational stream of an incompressible fluid. The wake is
composed of one or more recirculating regions, in each of which the total
pressure and the vorticity are uniform. Vortex sheets bound the various
regions. The behaviour of the flow in the corners and cusps which form
between the vortex sheets and the body, and the matching of the flows in
adjacent regions, lead to some conclusions about possible inviscid flow
structures.

It is both a pleasure and an honour to be here in Göttingen, one of the
main sources of inspiration and progress in fluid mechanics, to celebrate the
75th anniversary of the founding of the Aerodynamische Versuchsanstalt. It
is a particular privilege to bring you greetings from the Royal Aircraft
Establishment, another home of aeronautical research.

1   Introduction

The subject of our meeting is vortex flows. In aerodynamics, vortex
flows arise as boundary layers thicken and separate. It was Küchemann[1] who
stressed the important role of the shear in a thick boundary layer approaching the trailing edge of an aerofoil. He showed how a closed eddy of
reversed flow can arise, as a shear flow, with slip at the wall, approaches a
trailing edge of non-zero included angle. The reversed flow comes about, for
appropriate conditions, without the intervention of frictional forces.
Further details were worked out by Peter Smith[2]. It now appears that this
mechanism, by which the layer of fluid adjacent to the wall is retarded by
the vorticity in the essentially inviscid portion of the boundary layer
outside it, may play a vital part in the separation of a turbulent boundary
layer. This contrasts with the separation of a laminar layer, in which
Sychev's treatment indicates an essentially passive role for the upstream
boundary layer.

The present paper is concerned with the simulation of flows which have
separated, rather than with the process of separation. The result of separation is that the shear generated by viscous forces in the boundary layer is
convected into the body of the fluid. There, viscous forces play a negli-
gible part and inviscid models may be useful. In flows which are highly
three-dimensional, the occurrence of separation is not necessarily
associated with stagnation of the fluid. Appreciable convective velocities
remain, so that the transport of vorticity is effected predominantly by

convection, with diffusion playing a subordinate role. The shear is confined to the neighbourhood of the stream surfaces which spring from the
separation lines. When separation takes place only at salient edges, a
completely inviscid treatment is possible. There are now many such treat-
ments of the classical flow past a delta wing, on which Mangler and I worked
many years ago[3].

When separation takes place from a line on a smooth surface, the
inviscid model of the separated flow must be supplemented by a viscous
treatment of the separation process itself, if the position of the separation
line is to be determined. When the boundary layer is laminar at separation,
the account of the separation process proposed by Sychev[4] and confirmed by
Frank Smith[5] has been shown to be applicable in three-dimensional flow[6].
Using this work, Fiddes[7] has calculated the flow past slender cones at
angles of incidence large enough for separation to occur and give rise to
vortices. His predictions of the dependence of the position of laminar
separation on Reynolds number agree quite well with the water-tunnel observations of Rainbird *et al.*[8] Those calculations are for flow with lateral
symmetry. It is known that when the angle of incidence is two or more times
the semi-angle of a circular cone, vortices are sometimes observed in positions which are not symmetrically disposed about the incidence plane. Such
flows give rise to lateral forces and moments which may be of either sign,
are hard to measure with certainty, and are troublesome to allow for in
design. A vortex-sheet model exhibiting this behaviour has recently been
described[9]. It is of interest today to note that the RAE theoretical work
on this problem is proceeding in conjunction with experimental studies here,
at Göttingen, under the auspices of the Group for Aeronautical Research and
Technology in Europe.

The feature which makes such inviscid models of three-dimensional
separated flow useful is the dominance of convection and, in principle,
vortex stretching, over diffusion. The same dominance prevails in planar
separated flows which are sufficiently unsteady. For instance, in the
dynamic stall of a helicopter rotor, a separation vortex is formed, grows,
and is shed downstream on every passage of a retreating blade. Problems of
this kind will yield to inviscid modelling as soon as an adequate description
of the separation process itself is available.

On the other hand, when we come to what is usually thought of as the
simplest type of flow - steady, planar flow with constant density - the
situation is quite different. Planar separation typically involves stagnation, recirculation, and reattachment; the formation of a bubble with closed
streamlines. Here the velocity is low and diffusion plays a vital part in
establishing the steady flow pattern. On the other hand, compressibility
plays a minor role, and local solutions based on the constant density
approximation may have wider significance. Even when the global Reynolds
number is high, small frictional forces acting for a long time on the fluid
in a recirculating region reduce the total pressure below its free-stream
value. Moreover, since the separated region is closed, circulation cannot
be convected out of it; and so, since circulation is diffused into it from
the shear layer, we must expect to find that the flow in the separated region
is rotational. Since we are thinking of high Reynolds numbers, the vorticity
in the steady limit will be constant along each streamline. Furthermore, in
accordance with the principle enunciated by Prandtl[10] and elaborated by
Batchelor,[11,12] the vorticity will actually be uniform in the whole of the
region. More precisely, it will be uniform in each region of closed streamlines - of which there may be more than one in a single separated region.

In the classical inviscid model of a separated region, the vorticity is zero, there is no motion, the pressure is uniform, and we call it a dead-water region. For a closed region in particular this may well be needlessly restrictive. If we can find inviscid solutions containing separated regions in which the vorticity is piece-wise uniform, these will provide possible alternative limiting forms of the flow appropriate to indefinitely large Reynolds number. We cannot expect that inviscid considerations will determine the magnitude of the vorticity. On the other hand, if the flow is regarded as the limit of a sequence of viscous flows in which the Reynolds number tends to infinity, the vorticity will not be arbitrary. In some cases it has already been shown[13,14] how the vorticity can be determined from a consideration of the boundary layer surrounding the rotational region. However, these cases are of rotational regions with smoothly curved boundaries, so that a single, periodic, boundary layer forms on the entire boundary. A separated region on a solid surface has cusps or corners, as we shall see, and presents a harder problem.

In the present paper neither the determination of the vorticity nor the construction of the global inviscid solution is considered. Instead, attention is devoted to the behaviour of inviscid solutions close to the points on the solid surface from which separation is taking place. This means, in particular, that no contribution will be made to the question raised by Gersten et $al.$[15] about the possible inviscid, rotational solutions for cavity flow.

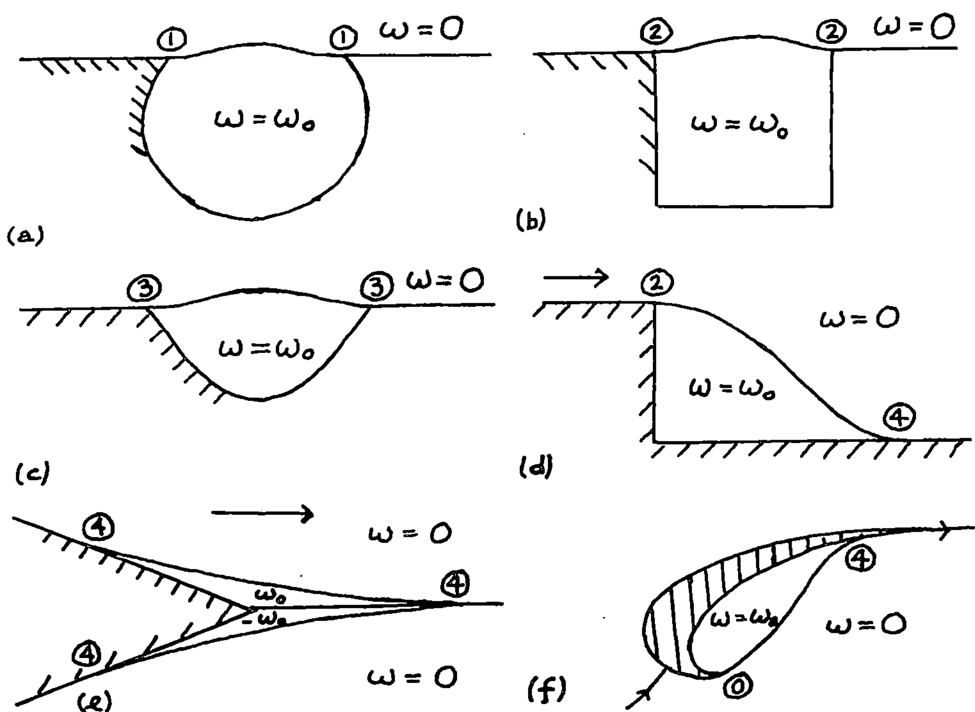

Sketches (a)-(f) show a number of simple geometries for which a
Prandtl-Batchelor model of the flow might be appropriate. (a), (b) and (c)
show symmetrical sharp-edged cavities, (d) shows flow down a step, (e) shows
a symmetrical trailing edge flow, and (f) shows a leading-edge slat, which is
'ideal', in the sense that it can retract smoothly onto a smooth aerofoil.
The points numbered ⓪-④ form a simple sequence. In each case two flows,
one rotational and one irrotational, are separated by a vortex sheet and by a
wedge of included angle δ. The angle δ increases from zero at point ⓪,
through $\pi/2$ at point ②, to π at point ④.

As a first step, the known solutions for flows in corners are recalled
and flow in a cusp is considered. These solutions are then matched across a
vortex sheet in order to produce the local solutions corresponding to points
⓪-④ in the sketches. Finally, various distortions of the trailing edge
flow (e), arising from departures from symmetry, are briefly explored.

2   Corner and cusp flow

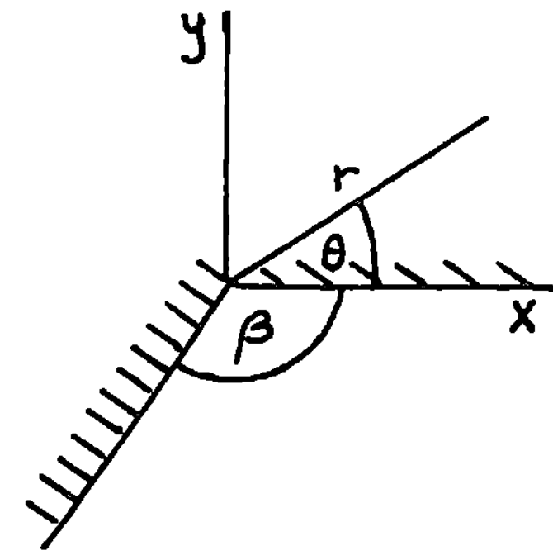

(g)

We consider first flow between two streamlines meeting at a non-zero
angle. Sufficiently close to the corner we assume the flow is the same as it
would be if the streamlines were straight and met at the same angle, β. We
want to consider flows with uniform vorticity, $\omega = \omega_0$. However, such a flow
is indeterminate in the sense that any multiple of the potential flow in the
corner may be added to it without changing either the vorticity or the
boundary conditions on the straight streamlines. It is therefore relevant to
recall first the potential flow in the corner.

For the configuration of sketch (g), the complex potential, W, is given
by

$$W \;=\; \phi + i\psi \;=\; Az^{\pi/\beta} \;, \qquad A \text{ real}, \quad z \;=\; x + iy \;=\; re^{i\theta} \;,$$

since this makes $\psi = 0$ for $\theta = 0$ and $\theta = -\beta$. The complex conjugate of the
velocity vector follows at once:

$$u - iv \;=\; \frac{dW}{dz} \;=\; \frac{\pi A}{\beta} z^{(\pi-\beta)/\beta} \;.$$

The occurrence of an infinite velocity at the corner is associated with a
value of β greater than π. Since only one angle greater than π can arise at
a point, no matching between the flows in different sectors is possible if
one of the angles exceeds π. We therefore take $\beta \leqslant \pi$ in what follows.
Since we shall be concerned with matching flows across the boundary streamlines, the quantity of primary interest is the tangential velocity component,
v_t. For the potential flow in a corner of angle β:

160

$$v_t \sim \frac{\pi A}{\beta} r^{(\pi-\beta)/\beta} \quad . \tag{1}$$

The solutions for rotational flow in a sector with straight walls may be found in a paper by Fraenkel[16]. Provided $\beta < \pi/2$, the behaviour of the tangential velocity component on the wall, v_t, sufficiently close to the corner, is given by

$$v_t \sim \pm\tfrac{1}{2}\omega_0 r \tan \beta \quad , \tag{2}$$

where ω_0 is the uniform vorticity. The unique asymptotic behaviour arises because, as $r \to 0$, (2) dominates (1) for $\beta < \pi/2$. For $\beta > \pi/2$, a solution of the same form arises for the rotational flow, but the potential flow dominates it, and equation (1) gives the asymptotic behaviour of v_t. When $\beta = \pi/2$, Fraenkel combines the rotational solution with a specific multiple of the potential solution to obtain a finite result. The asymptotic behaviour of the tangential velocity for $\beta = \pi/2$ is

$$v_t \sim \pm(2\omega_0/\pi) r \ln r \quad . \tag{3}$$

In summary, as β increases from zero, the dominant behaviour of v_t is initially locally determined, with v_t proportional to r. The coefficient of proportionality increases with β, and tends to infinity when $\beta \to \pi/2$. For $\beta = \pi/2$, the dominant behaviour of the velocity is still locally determined, but the variation with r is more rapid. For $\beta > \pi/2$, only the form of the velocity is locally determined, with an exponent which falls from unity towards zero. This implies a rapid variation of the velocity, becoming steeper as $\beta \to \pi$. The rapid variation disappears when $\beta = \pi$ and the flow can be regular.

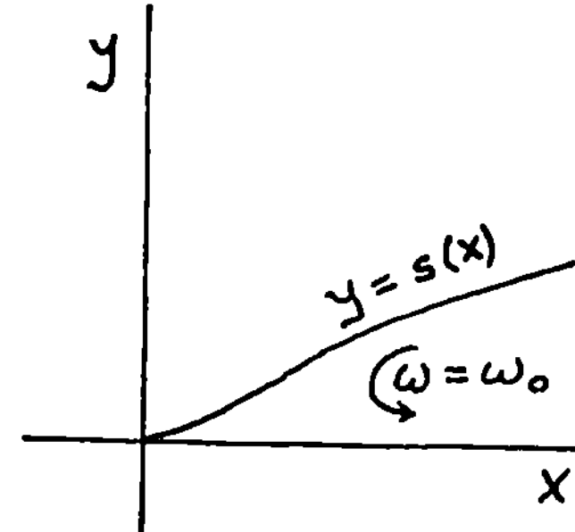

(h)

We now consider the flow between two streamlines which meet at a cusp, so that the angle β is zero. In line with the above discussion, we should expect the rotational contribution to be locally determined and to dominate the potential contribution. The rotational contribution is readily obtained using the slender eddy approximation introduced by Childress[17].

In the configuration of sketch (h), where fluid of uniform vorticity ω_0 occupies the region above the positive x-axis and below the curve $y = s(x)$, the key assumption is that y derivatives dominate x derivatives. The standard equation for the stream function, ψ, becomes

$$-\omega_0 = \nabla^2\psi \simeq \frac{\partial^2\psi}{\partial y^2} \quad .$$

The solution satisfying the boundary conditions $\psi = 0$ on $y = 0$ and $y = s(x)$ is

$$\psi = \tfrac{1}{2}\omega_0 y \left(s(x) - y\right) \quad .$$

161

The tangential velocity v_t is typified by $\partial\psi/\partial y$ on $y = 0$:

$$v_t = \tfrac{1}{2}\omega_0 s(x) \ . \tag{4}$$

Our confidence in this slender approximation is increased by noting the agreement between (2) and (4) when $s(x) \equiv x \tan \beta$.

We shall be interested in algebraic cusps, for which $s(x) \sim \lambda x^n$, $n > 1$. Equation (4) shows that the tangential velocity inside the cusped region then decreases towards zero at the stagnation point with just the same algebraic behaviour as the thickness of the region. Earlier work[18] has shown that the potential flow in such a region produces a tangential velocity on the boundary which decreases exponentially fast. This confirms our expectation that the rotational solution (4) dominates.

3   Flows with vortex sheets

It is perfectly possible for regions of flow with different uniform values of the vorticity, ω, to adjoin along streamlines across which the velocity is continuous. Unexpected singular types of behaviour may arise, as pointed out by Pierrehumbert[19]. Our present concern is with the behaviour of the flow past solid surfaces at high Reynolds number, so the vorticity originates in boundary layers, which extend into the fluid as shear layers. The distributed vorticity in the fluid arises by diffusion from the shear layers; so, in the limit of infinite Reynolds number, we must include a vortex sheet between a region of rotational flow and a region of potential flow. The value of the tangential velocity, v_t, will therefore in general be different on the two sides of the streamline which divides the flow.

Across such a vortex sheet, the pressure, p, must be continuous. Since viscous forces have operated to produce the uniform distribution of vorticity in the rotational flow, they will also have acted to reduce the total pressure, $p + \tfrac{1}{2}\rho\underline{v} \cdot \underline{v}$, in the rotational flow below its free-stream level. In particular, on the vortex sheet v_t^2 is smaller on the side of the rotational flow. However, the vortex sheet is a streamline of both flows, so the total pressure remains constant along each side of it, and so the difference in v_t^2 across it remains constant. For the present local arguments a convenient form of this is:

$$\frac{d}{d\sigma} \ (v_t^2) \quad \text{is continuous across the sheet} \ , \tag{5}$$

where σ is the arc length along the sheet.

We have been considering the dependence of the flow behaviour on the angle β between two streamlines of a flow with uniform vorticity. We now return to the role of the angle δ between the two streamlines forming the body at the point of separation or reattachment. For separation from a smooth wall, or reattachment to a smooth wall or wake, $\delta = \pi$. For separation from a cusped trailing edge, $\delta = 0$. For separation from, or reattachment to, a wedge-shaped trailing edge, $0 < \delta < \pi$. The possibility of separation from or reattachment to a corner which is concave to the flow can be excluded by a simple argument, so we do not consider $\delta > \pi$. Since our eventual interest is in engineering applications, we assume the wall shape is regular on either side of the point of interest.

Let us start with the smooth wall, $\delta = \pi$. For the present local argument this can be thought of as plane. The first point to establish is that the vortex sheet leaves the wall tangentially. Suppose that it did not, as in sketch (i). There must then be a stagnation point on either side.

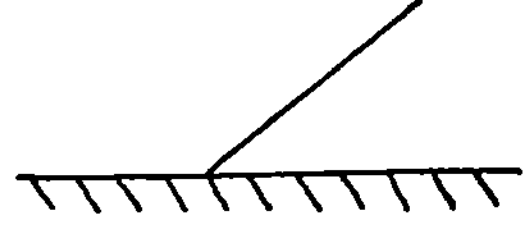

(i)

Since the pressure is continuous, the total pressure must be the same at each stagnation point. Hence, since the sheet is a streamline, the total pressure is continuous across the sheet everywhere. But the pressure, too, is continuous, so the tangential velocity is also continuous. There is therefore no vortex sheet, contradicting the assumption. Hence the sheet must leave tangentially, as in sketch (j).

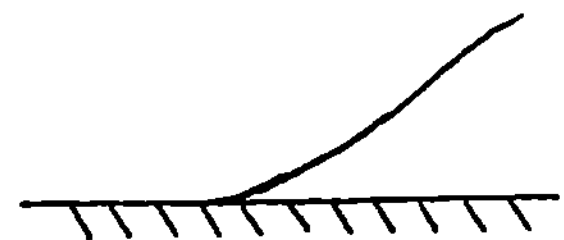

(j)

There is now a velocity U > 0 along the wall on the upstream side of the separation point, with a stagnation point on the downstream side. There is therefore a drop in total pressure of $\frac{1}{2}\rho U^2$ across the sheet, so we should not expect to find an irrotational, potential flow on the downstream side. One possibility is complete stagnation, with the vortex sheet forming a constant-pressure streamline of the outer flow. This possibility is exemplified in the familiar Kirchhoff solution for flow past a circular cylinder. In general, the curvature of the sheet tends to infinity at the separation point like $\sigma^{-\frac{1}{2}}$, where σ is measured from the separation point. This singular behaviour of the sheet is associated with a singularity of the same form in the pressure gradient on the wall upstream of separation. For particular positions of the separation point, this singular behaviour does not arise, leading to 'smooth separation'. This contrast between smooth and singular behaviour provides the inviscid background to the asymptotic theory of laminar separation for high Reynolds numbers of Sychev[4] and Smith[5].

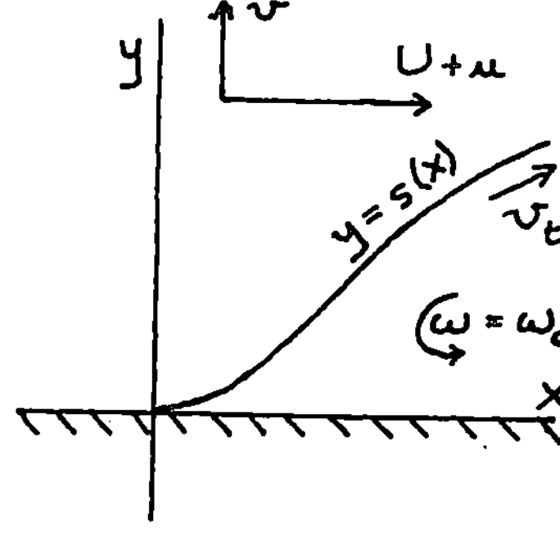

(k)

If there is flow on the downstream side, it will be rotational, and the tangential velocity, v_t, on the sheet will be given by (4). The flow above the sheet near the separation line can be described in a linearized approximation, with the sheet represented by a transpiration velocity on the wall. If the presence of the sheet introduces disturbance velocity

components u and v along and normal to the wall, the tangential velocity
component on the sheet is approximated locally by $U + u$, with $u = 0$ at the
separation point. It follows from (5) that, for $y = 0$ and $x > 0$,

$$2U \frac{\partial u}{\partial x} \sim \frac{d}{d\sigma} (v_t^2) \simeq \frac{d}{dx} (v_t^2) \ .$$

Hence, since both u and v_t vanish at the separation point

$$u \sim v_t^2/2U$$

and so, by (4)

$$u \sim \omega_0^2 s^2(x)/8U \qquad \text{for} \quad y = 0 \quad \text{and} \quad x > 0.$$

At the same time, since the wall and sheet are streamlines:

$$v \sim \begin{cases} Us'(x) & \text{for} \quad y = 0 \quad \text{and} \quad x > 0 \\ 0 & \text{for} \quad y = 0 \quad \text{and} \quad x < 0 \ . \end{cases}$$

These equations, coupled by the fact that u and $-v$ are the real and
imaginary parts of an analytic function of $x + iy$, represent a non-linear
integro-differential equation for $s(x)$. One family of solutions can be
constructed easily. If $s(x) \sim \lambda x^{N/2}$, $N = 1,2,3,\ldots,$ u becomes regular and
can be absorbed into the non-singular part of the flow. For a cusp, we need
$N > 1$. Then

$$v \sim \tfrac{1}{2}U\lambda N x^{\frac{1}{2}N-1} \ , \qquad N = 2, 3, \ldots$$

When the power of x in this relation is integral, logarithmic terms appear in
the conjugate function of v (see Ref.18 for details), which are absent from
the expression for u, so the corresponding, even, values of N must be dis-
carded. On the other hand, the half-power behaviour of v, given by odd
values of N, leads to a regular behaviour of its conjugate function, compat-
ible with a regular behaviour of u. A possible family of solutions is
therefore

$$s(x) \sim \lambda x^{N/2} \ , \qquad N = 3, 5, 7, \ldots \tag{6}$$

This behaviour is just that of the Kirchhoff free-streamline solutions,
so allowing the possibility of a rotational flow downstream of separation
does not change the analytic behaviour of the upstream flow. The numerical
magnitudes will, of course, be different. The framework of the Sychev-Smith
theory of laminar separation therefore carries over unchanged, but the predicted separation point will depend on the vorticity in the region of
separated flow. The constant pressure downstream which results from the
stagnant wake of the Kirchhoff flow is replaced by a slowly-varying pressure.
The variation across the separated region is too small to be reliably predicted by the slender eddy approximation. The streamwise variation is represented by that along the wall or along the sheet:

$$p - p_s = -\tfrac{1}{2}\rho v_t^2 = -\rho\omega_0^2 s^2(x)/8 = -\rho\omega_0^2\lambda^2 x^N/8 \ , \qquad N = 3, 5, \ldots \tag{7}$$

where p_s is the pressure at the separation point.

Coming now to the cusped trailing edge, $\delta = 0$, we find the presence of
rotation has no local effect. The separation streamline leaving the cusp
must leave tangentially, to avoid a sector of the fluid with angle $\beta > \pi$ and
a consequential unmatchable pressure singularity. Allowing the total pressure to be different on the two sides of the separation streamline means
that there will be a difference of tangential velocity across it, so that it
is a proper vortex sheet. However, each of the sectors of fluid at the

cusped edge has its angle $\beta = \pi$, so that the irrotational motion is dominant. For the most obvious case, in which the cusp is formed by two curves of different, finite, curvature, there must be a difference in curvature between the wall and the sheet, at least on one side. This curvature discontinuity gives rise to a logarithmic singularity in the gradient of the tangential velocity, which must be matched by a similar singularity in the flow on the other side of the sheet. Consequently, the curvature of the sheet will be intermediate between the curvatures of the two surfaces of the cusped trailing edge. Its value will be determined by equating the strengths of the singularity in $dv_t^2/d\sigma$ on the two sides, A and B:

$$v_A^2 (\Delta\kappa)_A = v_B^2 (\Delta\kappa)_B \quad ,$$

where V is the velocity and $\Delta\kappa$ is the jump in curvature. Hence

$$\kappa_s = \frac{v_A^2 \kappa_A + v_B^2 \kappa_B}{v_A^2 + v_B^2} \quad , \tag{8}$$

where κ_s is the curvature of the sheet and κ_A and κ_B are the curvatures of the two faces of the cusp.

The two limiting cases, $\delta = 0$ and π, have now been dealt with, leaving the general case of the salient edge with non-zero angle, $0 < \delta < \pi$, to be treated. Since $\delta > 0$, at least one of the angles β of the fluid sectors must be less than π, so that the fluid in it stagnates. For a non-trivial vortex sheet, the fluid in the other sector must not stagnate. Since fluid sector angles greater than π are excluded, the angle of the second sector must be π. The configuration is therefore as in sketch (ℓ), with $\beta = \pi - \delta$. The fluid

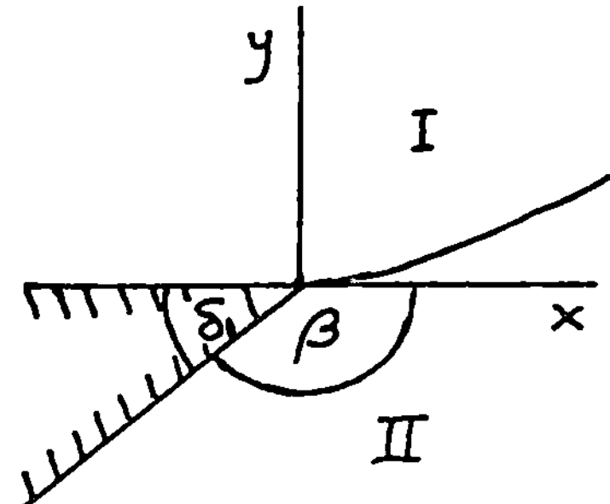

(ℓ)

in region II must have lower total pressure than that in region I in order to balance the static pressure across the vortex sheet at the edge. We therefore take the flow in region II to be rotational, with uniform vorticity ω_0, and the flow in region I to be potential. In the neighbourhood of the edge, the vortex sheet lies close to the positive real axis of the complex variable $z = x + iy$.

Following on from the case of the cusp, we consider first the range $0 \leq \delta < \pi/2$, as exemplified by sketch (ℓ). In accordance with equation (5), $dv_t^2/d\sigma$ must be continuous across the sheet. In region II, the flow near the edge will, in general, be dominated by the potential solution, since $\beta > \pi/2$. On this side, the tangential velocity on the sheet is therefore approximated by equation (1), in the form:

$$v_t \sim (\pi A/\beta) x^{\delta/\beta} \ .$$

In region I, the curvature of the sheet is relevant, but a linearised treatment should be adequate. If $U + u$ and v are the velocity components in this region parallel to the x and y axes, with $u = 0$ at the edge and U constant, then u and v will be small compared with U. The tangential velocity on this side is approximately $U + u$, so (5) requires

$$\frac{d}{dx} (U + u)^2 \sim \frac{d}{dx} \frac{\pi^2 A^2}{\beta^2} x^{2\delta/\beta} \ .$$

Neglecting higher order quantities, we find:

$$u = \lambda U x^{2\delta/\beta} \quad \text{for} \quad x > 0 \ , \quad \text{where} \quad \lambda = \pi^2 A^2/2\beta^2 U^2 \ . \tag{9a}$$

Since the wall is locally flat:

$$v = 0 \quad \text{for} \quad x < 0 \ . \tag{9b}$$

When $\beta = 2\pi/3$, $\delta = \pi - \beta = \beta/2$ and the expression (9a) is regular. In this case, no local singular behaviour arises, so that, in particular, the curvature of the sheet is continuous with the curvature of the wall. Further, the pressure gradient on both surfaces of the wall is finite. In region I, if the flow is off the edge, the gradient is favourable.

For any other value of β in the range $\pi/2 < \beta < \pi$, equations (9), together with the requirement that u and $-v$ are conjugate functions, define a local singular behaviour for the flow in region I. The singular behaviour can be shown to be unique by arguments like those of Ref.18, so it is enough to present a solution and verify that conditions (9) are satisfied. Consider first

$$u - iv = \lambda U \frac{e^{-im\pi}}{\cos m\pi} z^{m+1} \ , \quad \text{where} \quad m = \frac{2\delta - \beta}{\beta} \ . \tag{10}$$

If z is real and positive, $z = x > 0$, and $u = \lambda U x^{m+1}$, agreeing with (9a). If z is real and negative, $z = |x| e^{i\pi}$, $z^{m+1} = -|x|^{m+1} e^{im\pi}$, and so $v = 0$, agreeing with (9b). However, the solution (10) fails if $\cos m\pi = 0$, i.e. if $m = N + \tfrac{1}{2}$ for N integral. Consider therefore

$$u - iv = \lambda U z^{m+1} \left(1 + (i/\pi) \ln z\right) \ , \quad \text{for} \quad m = N + \tfrac{1}{2} \ . \tag{11}$$

If z is real and positive, $z = x > 0$, $u = \lambda U x^{m+1}$, agreeing with (9a). If z is real and negative, $z = |x| e^{i\pi}$, $\ln z = \ln |x| + i\pi$, and $z^{m+1} = -|x|^{m+1} e^{im\pi} = \pm i |x|^{m+1}$, so that $v = 0$, agreeing with (9b). The only values of N which lead to values of β in the range $\pi/2 < \beta < \pi$ are $N = 0$ and -1, for which $m = \pm\tfrac{1}{2}$, and $\beta = 4\pi/7$ and $4\pi/5$.

The slope of the sheet is given by v/U and so its curvature is

$$\kappa \sim \frac{1}{U} \frac{dv}{dx} \ , \quad \text{for} \quad y = 0, \ x > 0 \ .$$

Hence, for the general solution (10):

$$\kappa \sim \lambda (m + 1) \tan m\pi \ x^m \ . \tag{10a}$$

For the special solution (11)

$$\kappa \sim (\lambda/\pi) (m + 1) x^m \ln (1/x) \ . \tag{11a}$$

The pressure coefficient C_p on the wall upstream of the trailing edge (referred to the pressure at the edge) behaves like $-2u/U$ for $z = |x|e^{i\pi}$. For the general solution (10):

$$C_p \sim 2\lambda |x|^{m+1} \cos m\pi \ . \tag{10b}$$

For the special solution (11)

$$C_p \sim (-)^N (2\lambda/\pi) |x|^{m+1} \ln (1/|x|) \ . \tag{11b}$$

We can tabulate these results as follows:

δ	β	m	sheet curvature	wall pressure gradient
$\left(0,\ \dfrac{\pi}{5}\right]$	$\left(\pi,\ \dfrac{4\pi}{5}\right]$	$\left(-1,\ -\tfrac{1}{2}\right]$	infinite, positive	infinite, adverse
$\left(\dfrac{\pi}{5},\ \dfrac{\pi}{3}\right)$	$\left(\dfrac{4\pi}{5},\ \dfrac{2\pi}{3}\right)$	$\left(-\tfrac{1}{2},\ 0\right)$	infinite, negative	infinite, favourable
$\dfrac{\pi}{3}$	$\dfrac{2\pi}{3}$	0	continuous	finite, favourable
$\left(\dfrac{\pi}{3},\ \dfrac{3\pi}{7}\right)$	$\left(\dfrac{2\pi}{3},\ \dfrac{4\pi}{7}\right)$	$\left(0,\ \tfrac{1}{2}\right)$	continuous	zero, favourable
$\left[\dfrac{3\pi}{7},\ \dfrac{\pi}{2}\right)$	$\left[\dfrac{4\pi}{7},\ \dfrac{\pi}{2}\right)$	$\left[\tfrac{1}{2},\ 1\right)$	continuous	zero, adverse

For $\pi/3 < \delta < \pi/2$, the flow singularity does not affect the quantities of physical interest. For wedge angles less than $\pi/3$, both curvature and pressure gradient are infinite. For $\pi/5 < \delta < \pi/3$, the sheet is curved towards the region of reduced pressure and the wall pressure gradient is favourable.

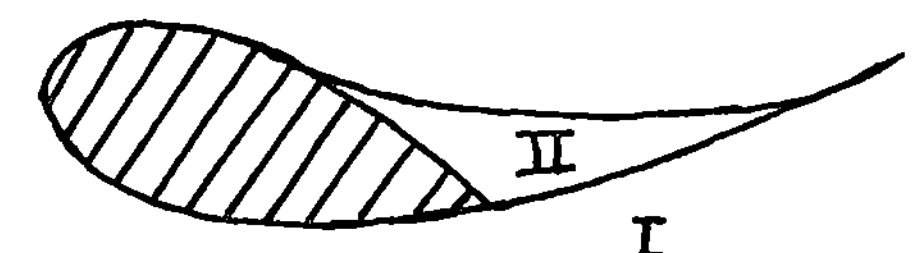

(m)

Hence, interpreting this as part of the flow past a very thick aerofoil, we might draw sketch (m). The upper surface flow separates to give a closed region of reduced total pressure, while the lower surface flow remains attached, with a favourable gradient at the trailing edge. We shall see later that this picture may need modification.

For $\delta < \pi/5$ (= 36°), corresponding to a realistic aerofoil, the vortex sheet initially curves away from the separated region. The pressure gradient at the trailing edge is infinitely adverse. A laminar boundary layer would probably separate upstream of the trailing edge, but the turbulent layer occurring in aeronautical practice might be less drastically affected.

We return later to consider trailing edge flows. For the moment, it is necessary to point out that the curious behaviour for small wedge angles arises from the occurrence of the potential flow singularity (1) in region II. Presumably if the shape of region II were to satisfy some global condition, the coefficient of the singularity in the potential flow would vanish, leaving the rotational flow, with the regular behaviour (2).

We now need to consider the case of the vortex sheet leaving a wedge

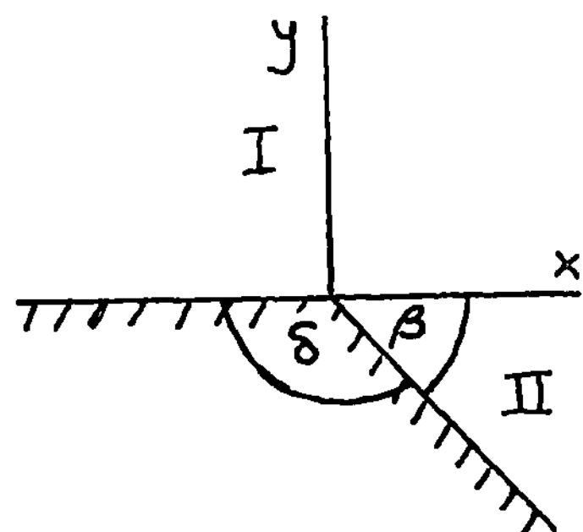

(n)

of angle δ greater than $\pi/2$, so that $\beta < \pi/2$. The flow in the sector of angle β is now dominated by the rotational contribution (2):

$$v_t \sim \pm \tfrac{1}{2}\omega_0 x \tan \beta \quad .$$

With U, u and v as before, condition (5) can be expressed as

$$\frac{d}{dx} (U + u)^2 \sim \frac{d}{dx} \left(\frac{\omega_0^2 \tan^2 \beta}{4} x^2 \right) \quad ,$$

so that

$$u \sim \left(\omega_0^2 \tan^2 \beta / 8U \right) x^2 \quad , \qquad \text{for} \quad x > 0 \ . \qquad (12)$$

This is a regular behaviour, which can be produced as part of the global solution. The sheet should therefore leave the edge with the same curvature as the wall. The upstream disturbance velocity will then have the same form as (12), implying a pressure gradient which is adverse but falling to zero at the edge itself, as in the last line of the table above.

Finally, there is the case of the right-angled corner, $\delta = \beta = \pi/2$. By (3) we have

$$v_t \sim \pm (2\omega_0/\pi) x \ln x$$

and the same argument as before leads to the specification on the real axis, $y = 0$, of

$$u \sim \mu U (x \ln x)^2 \ , \qquad \mu = 2\omega_0^2/\pi U^2 > 0 \ , \qquad \text{for} \quad x > 0 \ ,$$

and

$$v \sim 0 \qquad \qquad \text{for} \quad x < 0 \ .$$

An analytic function with this behaviour is

$$u - iv = \mu U z^2 (\ln z - i\pi)^2 \quad ,$$

as can easily be verified. The slope of the sheet is given by

$$\frac{dy}{dx} = \frac{v}{U} \sim 2\pi\mu x^2 \ln x \quad , \qquad \text{for} \quad x > 0 \ , \qquad (13)$$

so that there is no singularity in the curvature. Similarly, the pressure coefficient on the wall, referred to conditions at the edge, is

$$c_p = - \frac{2u}{U} = - 2\mu x^2 (\ln |x|)^2 \quad . \tag{14}$$

Consequently, the curvature of the sheet will be continuous with that of the wall and the pressure gradient will be zero at the edge and adverse just upstream. Thus $\delta = \pi/2$ also fits into the last line of the table above.

4 Reattachment and wake closure

The local flow solutions that have been considered are reversible, so the behaviour of reattaching flows is similar to the behaviour of separating flows, in this inviscid approximation. The pressure gradient, however, is reversed, because the flow direction is reversed, while the pressure remains the same.

One feature is that, where separation and reattachment involve stagnation of the flow in the same recirculating region, as in sketches (a)-(e), the pressure at reattachment must be equal to the pressure at separation. In this respect, allowing flow in the recirculating region gives no increase in generality over the Kirchhoff model. Any loss of total pressure along the vortex sheet which divides the regions is due to viscous effects which have been omitted from the present model.

For the symmetrical wake of sketches (e) and (o), there is no difficulty about local behaviour at wake closure. With a 3/2-power behaviour at B

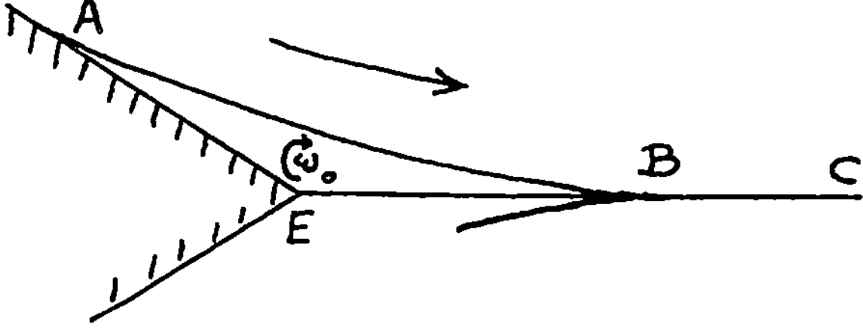

(o)

in the streamline AB, and a straight downstream continuation of it, there is a regular pressure rise as B is approached, with an infinitely favourable gradient on the downstream side of B. Return flows like that along BE have been discussed by Harper[20], who showed that multiple eddies are not necessarily produced, as had been conjectured.

For a small departure from the symmetrical condition, typically on an aerofoil at an angle of incidence small relative to the trailing edge angle, we should expect separation still to occur upstream of the trailing edge on the lower surface, though, in sketch (p), S_2 will be closer to the trailing

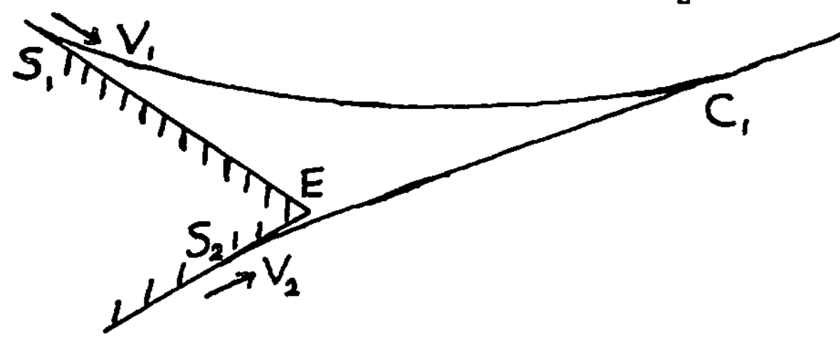

(p)

edge than S_1 is. Since the flow cannot negotiate the sharp edge at E, a
dividing streamline must run between E and the point C_1 of final downstream
closure. This streamline C_1E divides the wake into two recirculating regions
in which the sign of the vorticity will be different and the level of the
total pressure may be.

If the speeds V_1 and V_2 at the separation points S_1 and S_2 are equal,
the total pressure is the same in the two recirculating regions. The return-
ing dividing streamline leaves C_1 with a curvature intermediate between the
curvatures of the sheets S_1C_1 and S_2C_1 and approaches E along the bisector of
the trailing edge angle, as in sketch (q). The velocity is continuous
across C_1E.

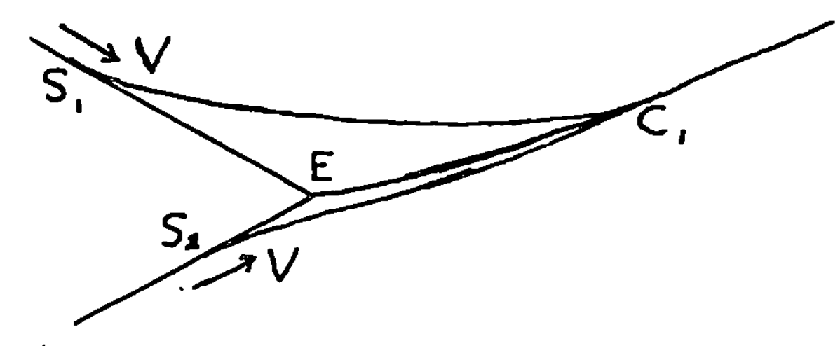

(q)

In the more general case of different speeds V_1 and V_2 at the separation points, the total pressure is different in the two recirculating regions.
As the dividing streamline leaves C_1 it initially follows one or other of the
vortex sheets C_1S_1 and C_1S_2, leaving it tangentially at a point C_2. It
approaches E along one of the tangents to the surface there. Sketch (r)

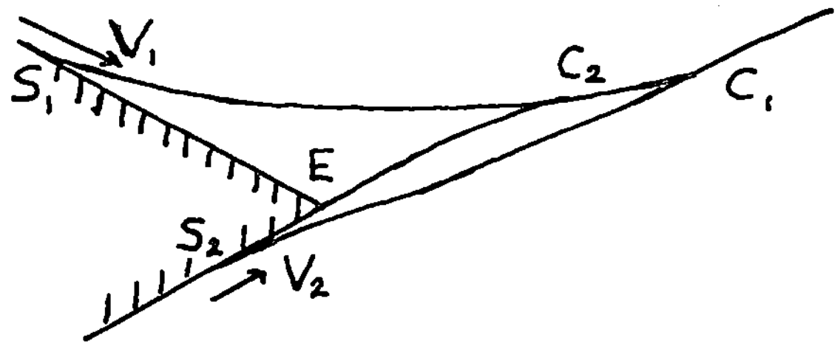

(r)

illustrates the configuration for the case $V_1 > V_2$, which corresponds to the
higher pressure on the lower surface of the aerofoil. An alternative to
sketch (r) is possible: C_2 might lie on C_1S_2, with the dividing streamline
leaving C_2 tangentially to C_1S_2, but towards C_1. It would then have to turn
through 180° to reach E. If $V_1 < V_2$, the dividing streamline is tangential
to the upper surface of the aerofoil. In the structure corresponding to
sketch (r), C_2 then lies on C_1S_2, but again there is a second possibility
involving a change in the direction of the dividing streamline. In any case,
the dividing streamline C_2E is a vortex sheet. Inside the wake, there are
stagnation points at E and C_2 on one side of the sheet; on the other side of
the sheet, the speeds are equal and are given by $|V_1^2 - V_2^2|^{\frac{1}{2}}$.

For a larger departure from the symmetrical condition, we should
expect that separation on the lower surface would be delayed until the trailing edge is reached. The arguments set out in section 3 cast some doubt on
whether a laminar boundary layer would reach the trailing edge if the trailing edge angle is small, but we set these doubts aside for the present.
There must then be a vortex sheet leaving the trailing edge tangentially to
the lower surface. Suppose first that the speeds at separation are different,
$V_1 \neq V_2$. The total pressure in the wake is then different at S and E, so a

170

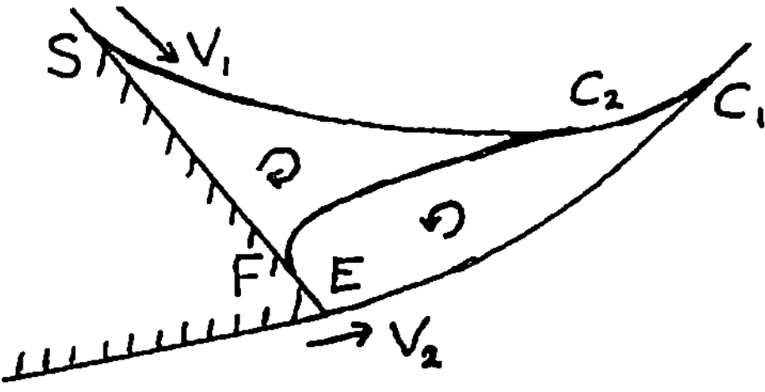

(s)

dividing streamline must separate the two regions of uniform total head. This dividing streamline must be tangential to SE, at a point F, say, and tangential to either SC_1 or EC_1 at a point C_2. If $V_1 > V_2$, the configuration of sketch (s) is the more plausible of the possibilities, though again C_2 might lie on EC_1. If $V_1 < V_2$, the orientation of the cusp on SE changes, and

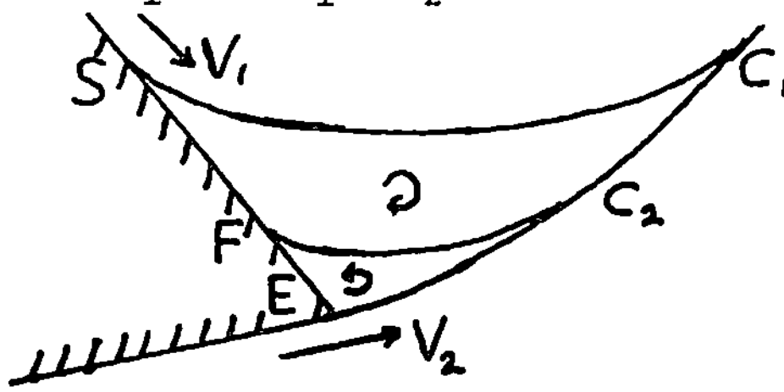

(t)

the more plausible pattern has C_2 on EC_1, as in sketch (t). In both cases, C_2F is a vortex sheet and the non-zero velocities in the wake at C_2 and F must be equal.

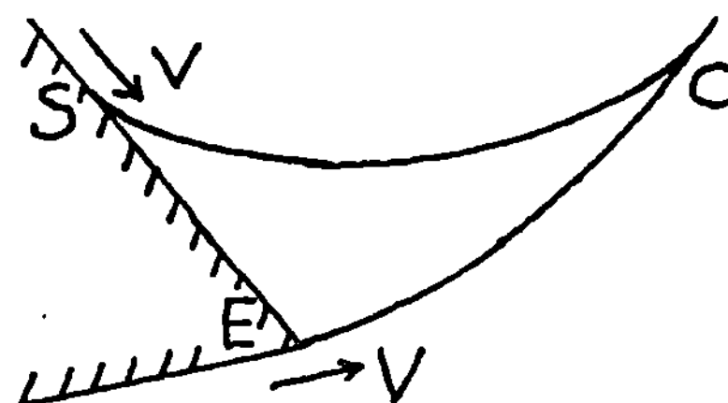

(u)

If $V_1 = V_2$, the total pressure is uniform and there is no necessity for a streamline dividing the wake into two regions of different vorticity - see sketch (u). If there is a dividing streamline, it must start from C with a curvature intermediate between those of CS and CE. If it were to meet SE between S and E, it would have to do so at right angles, since the pressure on the two sides must be equal. Even then, equality of pressure demands that the magnitude of the vorticity is the same on the two sides. This seems very restrictive. Otherwise, the dividing streamline must terminate at S or, more probably, at E, as in sketch (v). Its direction at E is determined by the

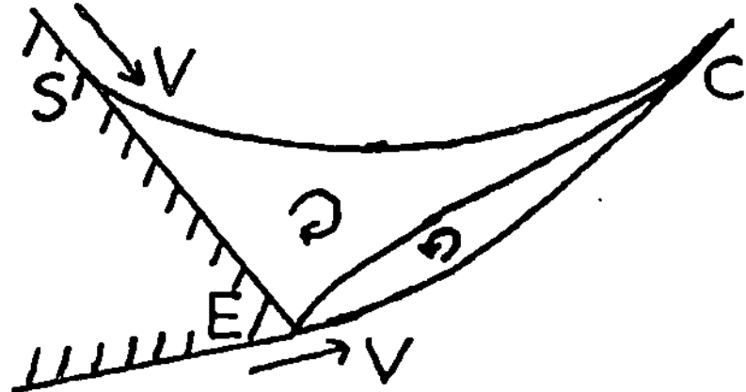

(v)

values of the vorticity on either side of it.

The flow patterns of sketches (q), (r), (s), (t) and (v) all show a
reattachment streamline proceeding upstream from the wake closure to the
vicinity of the trailing edge. Geometrical constraints force the region of
anticlockwise vorticity to be flat. In both respects these inviscid patterns
resemble solutions derived for high Reynolds number viscous flow by
F.T. Smith[21].

Finally, despite the number of apparently possible flow structures
which have been described, it must be admitted that many more are possible.
A difference in the speeds at the points of separation on the aerofoil makes
it necessary to introduce a second recirculating region. The resulting
structure then requires the equality of two other fluid speeds. There may be
enough freedom in the configuration to make them equal, but if there is not,
a third region with a third level of total pressure must be introduced. The
way in which an infinite sequence of such regions might be arranged is clear
from sketch (t). The present local arguments cannot resolve questions of
this sort. Careful calculation is required, and this must respect the rather
complex local analytic behaviour which has been shown to be necessary.

<u>References</u>

1   D. Küchemann, Zeitschrift für Flugwissenschaften $\underline{15}$, pp.292-294.

2   P.D. Smith, RAE unpublished, 1970.

3   K.W. Mangler & J.H.B. Smith, Proc. R. Soc. Lond. A $\underline{251}$, pp.200-217, 1959.

4   V.V. Sychev, Izv. Akad. Nauk., Mekh. Zhid. i Gaza, No.3, pp.47-59, 1972.

5   F.T. Smith, Proc. R. Soc. Lond. A $\underline{356}$, pp.443-463, 1977.

6   F.T. Smith, RAE TR 78095, 1978.

7   S.P. Fiddes, AGARD Symp. "Computation of Viscous-Inviscid Interactions',
    Colorado Springs, 1980, AGARD-CP-291.

8   W.J. Rainbird, R.S. Crabbe & L.S. Jurewicz, NRC (Canada) Aeronautical
    Report LR 385, 1963.

9   S.P. Fiddes & J.H.B. Smith, AGARD Symp. 'Missile Aerodynamics',
    Trondheim, 1982.

10 L. Prandtl, Ver. III Int. Math. Kongr., Heidelberg, 1904, pp.484-491;
    Ges. Abhand. $\underline{2}$, pp.575-584.

11 G.K. Batchelor, J. Fluid Mech. $\underline{1}$, 2, pp.177-190, 1956.

12 G.K. Batchelor, J. Fluid Mech. $\underline{1}$, 4, pp.388-398, 1956.

13 W.W. Wood, J. Fluid Mech. $\underline{2}$, 1, pp.77-87, 1957.

14 N. Riley, J. Eng. Math. $\underline{15}$, 1, pp.15-27, 1981.

15 K. Gersten, H. Herwig & P. Wauschkuhn, AGARD Symp. 'Computation of
    Viscous-Inviscid Interactions', Colorado Springs, 1980, AGARD-CP-291

16 L.E. Fraenkel, J. Fluid Mech. $\underline{11}$, 3, pp.400-406, 1961.

17 S. Childress, Phys. Fluids $\underline{9}$, 5, pp.860-872, 1966.

18 J.H.B. Smith, RAE TR 77058, 1977.

19 R.T. Pierrehumbert, J. Fluid Mech. $\underline{102}$, p.478, 1981.

20 J.F. Harper, J. Fluid Mech. $\underline{17}$, 1, pp.141-153, 1963.

21 F.T. Smith, United Tech. Res. Center Rep. UTRC 82-13, 1982.

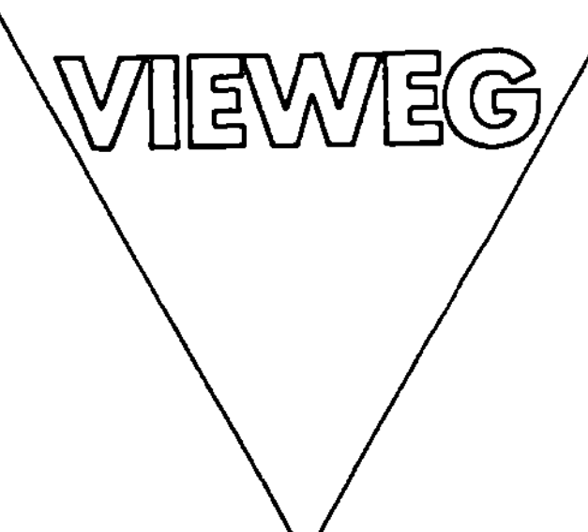

Notes on Numerical Fluid Mechanics

Volume 4
Ernst Heinrich Hirschel and Wilhelm Kordulla
Shear Flow in Surface-Oriented Coordinates
With 39 figures. 1981. X, 266 pages. 16,2 X 22,9 cm. Hardcover

Volume 5
Henri Viviand (Ed.)
Proceedings of the fourth GAMM-Conference on Numerical Methods in Fluid Mechanics
With 167 figures. 1982. VII, 343 pages. DIN C 5. Hardcover

Volume 6
Norbert Peters and Jürgen Warnatz (Eds.)
Numerical Methods in Laminar Flame Propagation
With 66 figures. 1982. VIII, 202 pages. DIN C 5. Hardcover

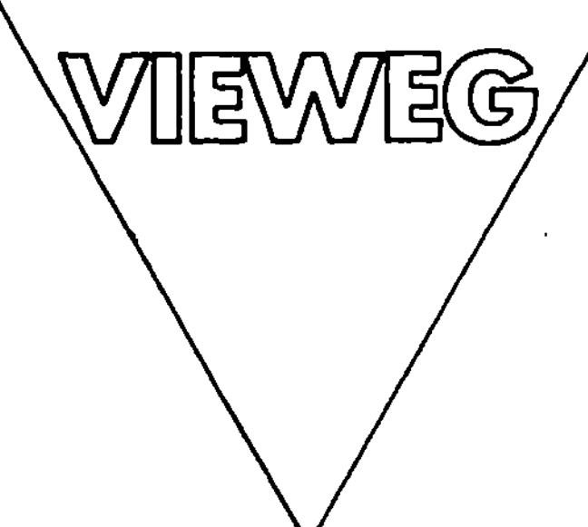

Notes on Numerical Fluid Mechanics

Volume 1
Karl Förster (Ed.)
Boundary Algorithms for Multidimensional Inviscid Hyperbolic Flows
With 92 figures. 1978. IV, 128 pages. 16,2 X 22,9 cm.
Hardcover

Volume 2
Ernst Heinrich Hirschel (Ed.)
Proceedings of the Third GAMM-Conference on Numerical Methods in Fluid Mechanics
With 157 figures. 1980. VII, 315 pages. 16,2 X 22,9 cm.
Hardcover

Volume 3
Arthur Rizzi and Henri Viviand (Eds.)
Numerical Methods for the Computation of Inviscid Transonic Flows with Shock Waves
A GAMM-Workshop. With 121 figures. 1981. XVI, 280 pages.
16,2 X 22,9 cm. Hardcover